湖南大学建筑与规划学院教学成果丛书

设计的生成 过程与教学

湖南大学建筑与规划学院优秀课程设计汇编

2015-2021

Compilation of Curriculum Design of School of
Architecture and Planning, Hunan University

湖南大学建筑与规划学院教学成果编写组 编

中国建筑工业出版社

图书在版编目（CIP）数据

设计的生成　过程与教学：湖南大学建筑与规划学院优秀课程设计汇编：2015-2021 = Compilation of Curriculum Design of School of Architecture and Planning, Hunan University / 湖南大学建筑与规划学院教学成果编写组编. -- 北京：中国建筑工业出版社，2022.9

（湖南大学建筑与规划学院教学成果丛书）

ISBN 978-7-112-27761-2

Ⅰ．①设… Ⅱ．①湖… Ⅲ．①建筑学－课程设计－汇编－高等学校 Ⅳ．①TU-0

中国版本图书馆CIP数据核字(2022)第147482号

责任编辑：陈夕涛 李东 徐昌强
责任校对：王烨

湖南大学建筑与规划学院教学成果丛书

设计的生成　过程与教学
湖南大学建筑与规划学院优秀课程设计汇编
2015-2021
Compilation of Curriculum Design of School of
Architecture and Planning, Hunan University

湖南大学建筑与规划学院教学成果编写组　编

*

中国建筑工业出版社出版、发行（北京海淀三里河路9号）

各地新华书店、建筑书店经销

北京富诚彩色印刷有限公司印刷

*

开本：787毫米×1092毫米　1/16　印张：18　字数：565千字
2022年11月第一版　　2022年11月第一次印刷

定价：168.00元

ISBN 978-7-112-27761-2

（39018）

湖南大学建筑与规划学院教学成果丛书编委会

顾　问：魏春雨
主　编：徐　峰
副主编：袁朝晖　焦　胜　卢健松　叶　强　陈　翚　周　恺

设计的起点　认知与启蒙
湖南大学建筑与规划学院优秀基础教学成果汇编 2015-2021
执行主编：钟力力
参与编辑：胡梦倩　陈瑞琦　林煜芸　亓宣雯

设计的生成　过程与教学
湖南大学建筑与规划学院优秀课程设计汇编 2015-2021
执行主编：许昊皓
参与编辑：李　理　齐　靖　向　辉　邢书舟　王　蕾　刘　晴　刘　骞　杨赛尔　高美祥
　　　　　王　文　陆秋伶　谭依婷　燕良峰　尹兆升　吕潇洋

设计的检验　理性与创新
湖南大学建筑与规划学院优秀毕业设计汇编 2015-2021
执行主编：杨　涛　姜　敏
参与编辑：王小雨　黄龙颜　王　慧　李金株　张书瑜　闫志佳　叶　天　胡彭年

设计的实践　转译与传承
湖南大学建筑与规划学院优秀实践案例汇编 2015-2021
执行主编：沈　瑶
参与编辑：张　光　陈　娜　黎璟玉　廖静雯　林煜芸　陈瑞琦　陈偌晰　刘　颖　欧阳璐

设计的理论　在地与远方
湖南大学建筑与规划学院优秀研究论文汇编 2015-2021
执行主编：沈　瑶
参与编辑：何　成　冉　静　成逸凡　张源林　廖土杰　王禹廷

序言

千年学府，百年建筑，湖南大学是中国早期现代建筑人才培养的摇篮之一。

刘敦桢、蔡泽奉、柳士英诸位先贤，负笈东瀛，学成归国，1929 年，麓山之脚，湘江之滨，开启了建筑学科的办学。近百年来，又经由杨慎初、闵玉林、黄善言等诸位先生的传承，巫继光、蔡道馨、柳展辉等老师的延续，以及当下众多教师的共同努力，依托湖湘地域文化、经世致用哲学、地方实践演练，形成了富有地方特色的现代建筑人才培养体系。

90 余年的发展，在行业变化与学科分合中，湖南大学建筑学科逐渐成长。1962 年成为第一批国务院授权的建筑学专业硕士研究生招生单位；1984 年成立建筑系；1990 年恢复五年制本科；1996 年首次通过专业评估；2003 年成立"建筑学院"；2008 年及 2015 年本科及硕士研究生培养评估通过（有效期 7 年）。拥有建筑学以及学科博士学位授予权（2010）、建筑学博士后流动站（2014）、国家级一流本科专业建设点（2019）、地方建筑与技术国际科技创新合作基地（2019）、丘陵地区城乡人居环境科学湖南省重点实验室（2020）。2021 年，更名为"建筑与规划学院"。

目前，按照"形式与认知、空间与环境、建构与营造、技术与综合、创作与实践"五个主题，湖南大学建筑学本科的人才培养依托级有序进阶，逐级推进。目前已经形成以设计课为主干，文化（建筑历史、城市研究）技术（建筑环境、建筑结构）类课程相融合的一轴两翼的课程组织模式。并以"综合课程包"的形式统筹融合，形成"10 个特色课程包"，贯通不同的知识体系。

"形式与认知"系列课程的教学，是空间、形体、材料抽象建构课程的综合，从平面、色彩、立体三大构成课入手，使学生掌握现代设计训练的基本方法，并逐渐实现具象美学思维向抽象美学思维的转换。在学年末，以"实体建构"工作营为大综合训练，在"3×3×3"米的范围内实施特定空间的营建，实现尺度、空间、材料、构成训练的融合。

"空间与环境"系列课程，通过室内环境、校园环境、城市环境的逐级推进，帮助学生由简入繁地理解身体与空间、建筑与环境的关系。多年以来，以"微型营建"的课程包实施相应的教学内容，选定特定地段，指导学生设计小型建筑与小微聚落，在单体设计的基础上适度融入场地设计的内容。近年来，辅以"图形转化"课程设置，联系二维图形向三维空间的转变，提升学生兴趣，进一步推动形态抽象与空间塑造的能力提升。

"建构与营造"系列课程，在结构、构造课程的基础上，提升建筑设计课程中材料搭接、细部设计的能力；与此同时，强化场景的营建能力，注重城市文脉与场所关联。课程强调技术与人文的融合，以城市人文色彩浓厚的博物馆、展览馆等建筑类型的训练为主。

"技术与综合"阶段包含"绿色高层"与"数字大跨"两个"课程包"，统筹建筑设计与建筑性能模拟、空间结构创新、复杂场地设计、建筑生成设计等教学内容。以高层与大跨建筑为主要类型的课程设计，将融合城市设计、场地设计、暖通设备等教学内容，既是对前期教学内容的补遗，也最能综合体现现代建筑的难度与复杂度。

"创作与实践"阶段，将安排生产实习、毕业实习和毕业设计。为了鼓励同学们积极投身前沿实践，综合应用前面四年的学习成果，我们联合多个设计单位、科研院所作为实践基地，组织跨高校、跨学科的多种类型的联合毕业设计；2020 年开始，每年组织题为"鹿鸣赫曦"的全国建筑类大学生毕业设计分享。

建筑学的本科学习，是一门重视技能培养的博雅教育。在湖南大学的课堂里，通过五年有序进阶的教育，通过创新力、协同力、实践力的培养，同学们将成长为在建筑学领域具有一定思辨能力与实践能力的专业人才，为其成长为行业内的中坚力量与领军人物奠定专业的基础。

湖南大学建筑与规划学院副院长

卢健松

总体介绍

学校概况

湖南大学办学历史悠久、教育传统优良，是教育部直属全国重点大学，国家"211 工程""985 工程"重点建设高校，国家"世界一流大学"建设高校。湖南大学办学起源于公元 976 年创建的岳麓书院，始终保持着文化教育教学的连续性。1903 年改制为湖南高等学堂，1926 年定名为湖南大学。目前，学校建有 5 个国家级人才培养基地、4 个国家级实验教学示范中心、1 个国家级虚拟仿真实验教学中心、拥有 8 个国家级教学团队、6 个人才培养模式创新实验区；拥有国家重点实验室 2 个、国家工程技术研究中心 2 个、国家级国际合作基地 3 个、国家工程实验室 1 个；入选全国首批深化创新创业教育改革示范高校、全国创新创业典型经验高校、全国高校实践育人创新创业基地。

学院概况

湖南大学建筑与规划学院的办学历史可追溯到 1929 年，著名建筑学家刘敦桢、柳士英在湖南大学土木系中创办建筑组。90 余年以来，学院一直是我国建筑学专业高端人才培养基地。学院下设两个系、三个研究中心和两个省级科研平台，即建筑系、城乡规划系、地方建筑研究中心、建筑节能绿色建筑研究中心、建筑遗产保护研究中心、丘陵地区城乡人居环境科学湖南省重点实验室、湖南省地方建筑科学与技术国际科技创新合作基地。

办学历程

1929 年，著名建筑学家刘敦桢在湖南大学土木系中创办建筑组。

1934 年，中国第一个建筑学专业 —— 苏州工业专门学校建筑科的创始人柳士英来湖南大学主持建筑学专业。柳士英在兼任土木系主任的同时坚持建筑学专业教育。

1953 年，全国院系调整，湖南大学合并了中南地区各院校的土木、建筑方面的学科专业，改名"中南土木建筑学院"，下设营建系。柳士英担任中南土建学院院长。

1962 年，柳士英先生开始招收建筑学专业研究生，湖南大学成为国务院授权的国内第一批建筑学研究生招生院校之一。

1978 年，在土木系中恢复"文革"中停办的建筑学专业，1984 年独立为建筑系。

1986 年，开始招收城市规划方向硕士研究生。

1995 年，在湖南省内第一个设立五年制城市规划本科专业。

1996 年至 2004 年间，三次通过建设部组织的建筑学专业本科及研究生教育评估。

2005 年，学校改建筑系为建筑学院，下设建筑、城市规划、环境艺术 3 个系，建筑历史与理论、建筑技术 2 个研究中心和 1 个实验中心。2005 年，申报建筑设计及其理论博士点，获得批准。同年获得建筑学一级学科硕士点授予权。

2006 年设立景观设计系，2006 年成立湖南大学城市建筑研究所，2007 年成立湖南大学村落文化研究所。

2008 年，城市规划本科专业在湖南省内率先通过全国高等学校城市规划专业教育评估。

2010 年 12 月，获得建筑学一级学科博士点授予权，下设建筑设计及理论、城市规划与理论、建筑历史及理论、建筑技术及理论、生态城市与绿色建筑五个二级学科方向。

2010 年，将"城市规划系"改为"城乡规划系"。

2011 年，建筑学一级学科对应调整，申报并获得城乡规划学一级学科博士点授予权。

2012 年，城乡规划学本科（五年制）、硕士研究生教育通过专业教育评估。

2012 年，获得城市规划专业硕士授权点。

2012 年，教育部公布的全国一级学科排名中，湖南大学城乡规划学一级学科为第 15 位。

2014 年，设立建筑学博士后流动站。

2016 年，城乡规划学硕士研究生教育专业评估复评通过，有效期 6 年。

2017 年，在第四轮学科评估中为 B 类（并列 11 位）。

2019 年，建筑学专业获批国家级一流本科专业建设点，建成湖南省地方建筑科学与技术国际科技创新合作基地;

2020 年，城乡规划专业获评国家级一流本科专业建设点。

2020 年，建成丘陵地区城乡人居环境科学湖南省重点实验室。

2021 年，"建筑学院"更名为"建筑与规划学院"。

建筑学专业介绍

一、学科基本情况

本学科办学 90 余年以来，一直是我国建筑学专业的高端人才培养基地。1929 年，著名建筑学家刘敦桢、柳士英在湖南大学土木系中创办建筑组；1953 年改为"中南土木建筑学院"，成为江南最强的土建类学科；1962 年成为国务院授权第一批建筑学专业硕士研究生招生单位；1996 年首次通过专业评估以来，本科及硕士研究生培养多次获"优秀"通过；2011 年获批建筑学一级学科博士授予权；2014 年获批建筑学博士后流动站；2019 年获批国家级一流本科专业建设点。

二、学科方向与优势特色

下设建筑设计及理论、建筑历史与理论、建筑技术科学、城市设计理论与方法 4 个主要方向，通过科研项目和社会实践，实现前沿领域对接，已形成了"地方建筑创作""可持续建筑技术""绿色宜居村镇""建筑遗产数字保护技术"等特色与优势方向。

三、人才培养目标

承岳麓书院千年文脉，续中南土木建筑学院学科基础，依湖南大学综合性学科背景，适应全球化趋势及技术变革特点，着力培养创新意识、文化内涵、工程实践能力兼融的建筑学行业领军人才。

城乡规划专业介绍

一、学科基本情况

本学科是全国较早开展规划教育的大学之一，具有完备的人才培养体系（本科、学术型／专业型硕士研究生、学术型／工程类博士、博士后），湖南省"双一流"建设重点学科。本科和研究生教育均已通过专业评估，有效期 6 年。

二、学科方向与优势特色

学位点下设城乡规划与设计、住房与社区建设规划、城乡生态环境与基础设施规划、城乡发展历史与遗产保护规划、区域发展与空间规划 5 个主要方向，通过科研项目和社会实践，实现前沿领域对接，已形成了城市空间结构、城市公共安全与健康、丘陵城市规划与设计、乡村规划、城市更新与社区营造等特色与优势方向。学科建有湖南省重点实验室"丘陵地区城乡人居环境科学"、与湖南省自然资源厅共建"湖南省国土空间规划研究中心"、与住房与城乡建设部合办"中国城乡建设与社区治理研究院"。

三、人才培养目标

学科聚焦世界前沿理论，面向国家重大需求，面向人民生命健康，服务国家和地方经济战略，承担国家级科研任务，产出高水平学术成果，提供高品质规划设计和咨询服务，在地方精准扶贫与乡村振兴工作中发挥作用，引领地方建设标准编制，推动专业学术组织发展。致力于培养基础扎实、视野开阔、德才兼备，具有良好人文素养、创新思维和探索精神的复合型高素质人才。

Introduction to Hunan University

Hunan University is an old and prestigious school with an excellent educational tradition. It is considered a National Key University by the Ministry of education, is integral to the national "211 Project" and "985 Project", and has been named a national "world-class university". Hunan University as it is today, originally known as Yuelu Academy, was founded in 976 and has continued to maintained the culture, education, and teaching for which it was so well known in the past. It was restructured into the university of higher education that exists today in 1903 and officially renamed Hunan University in 1926. The university has five national talent training bases, four national experimental teaching demonstration centers, one national virtual simulation experimental teaching center, eight national teaching teams, and six talent training mode innovation experimental areas. The school is also well equipped in terms of facilities, as it has two national key laboratories, two national engineering technology research centers, three national international cooperation bases, and one national engineering laboratory. It has also received many honors, as it is considered one of the top national demonstration universities for deepening innovation and entrepreneurship education reform, one of the top national universities with opportunities in innovation and entrepreneurship, and one of the top national universities' for practical education, innovation, and entrepreneurship.

School overview

The origin of the School of Architecture and Planning at Hunan University can be traced back to 1929, when famous architects Liu Dunzhen and Liu Shiying founded the architecture group as part of the Department of Civil Engineering. For more than 90 years, it has been a high-level talent training base for architecture in China. The school has two departments, three research centers, and two provincial scientific research platforms, namely, the Department of Architecture, the Department of Urban and Rural Planning, the Local Building Research Center, the Energy-saving Green Building Research Center, the Building Heritage Protection Research Center, the Hunan Provincial Key Laboratory of Urban and Rural Human Settlements and Environmental Science in Hilly Aeas, and the Hunan Provincial Local Science and Technology, International Scientific and Technological Innovation Cooperation Base.

Timeline of the University of Hunan's development

In 1929, the famous architect Liu Dunzhen founded the construction group within the Department of Civil Engineering at Hunan University. In 1934, Liu Shiying, the founder of the Architecture Department of the Suzhou Institute of Technology, which was the first one to provide major in architecture in China, came to Hunan University to preside over architecture major. Liu Shiying insisted on architectural education while concurrently serving as the director of the Department of Civil Engineering.

In 1953, with the adjustment of national colleges and departments, Hunan University merged their disciplines of civil engineering and architecture with various colleges and universities in central and southern China, forming a new institution that was renamed "Central and Southern Institute of Civil Engineering and Architecture". At this new institution, they set up a Department of Construction. Liu Shiying served as president of the Central South Civil Engineering College.

In 1962, Liu Shiying began to recruit postgraduates majoring in architecture. Hunan University became one of the first institutions authorized by the State Council to recruit postgraduates in architecture in China.

In 1978, Liu Shiying resumed providing the architecture major in the Department of Civil Engineering, which had been suspended during the Cultural Revolution. The Department of Architecture became independent in 1984.

In 1986, the University of Hunan began to recruit master's students to study urban planning.

In 1995, the first five-year official undergraduate major in urban planning was established in Hunan Province.

From 1996 to 2004, the university passed the undergraduate and graduate education evaluation of architecture organized by the

Ministry of Construction three times.

In 2005, the school changed its architecture department into an Architecture College, which included the three departments of architecture, urban planning, and environmental art design, two research centers for architectural history, theory, and architectural technology respectively, and one experimental center.In 2005, the university applied to provide a doctoral program of architectural design and theory, which was approved. In the same year, it was also granted the right to provide a master's degree in architecture.

In 2006, the Department of Landscape Design and the Institute of Urban Architecture at Hunan University were established.In 2007, the Institute of Village Culture at Hunan University was established.

In 2008, the undergraduate major of urban planning took the lead in passing the education evaluation for urban planning majors in national colleges and universities in Hunan Province.

In December 2010, Hunan University was granted the right to provide a doctoral program in the first-class discipline of architecture, with five second-class discipline directions, including: Architectural Design and Theory, Urban Planning and Theory, Architectural History and Theory, Architectural Technology and Theory, and Ecological City and Green Building Design.

In 2010, the "Urban Planning Department" was changed to the "Urban and Rural Planning Department".

In 2011, the university applied for and obtained the ability to transform the first-class discipline of architecture to provide the right to grant the doctoral program of the first-class discipline of urban and rural planning.

In 2012, the undergraduate (five-year) and master's degree in education in urban and rural planning passed the professional education evaluation.

In 2012, it obtained the authorization to provide a master's in urban planning.

In the national first-class discipline ranking released by the Ministry of Education in 2012, the first-class discipline of urban and rural planning of Hunan University ranked 15th overall.

In 2014, a post-doctoral mobile station for architecture was established.

In 2016, the degree program for a Master of Urban and Rural Planning was given a professional re-evaluation and passed, which is valid for another 6 years.

In 2017, the university was classified as Class B and tied for 11th place in the fourth round of discipline evaluation.

In 2019, the architecture specialty was approved as a National First-Class Undergraduate Specialty Construction Site and built into an international scientific and technological innovation and cooperation base of local building science and technology in Hunan Province.

In 2020, the major of urban and rural planning was rated as a national first-class undergraduate major construction point.

In 2020, the school began construction on the Hunan Key Laboratory of Urban and Rural Human Settlements and Environmental Science in Hilly Areas.

In 2021, the "School of Architecture" was renamed the "School of Architecture and Planning".

Introduction to architecture

1. Discipline overview

This university has provided a high-level talent training base for architecture in China for more than 90 years. In 1929, famous architects Liu Dunzhen and Liu Shiying founded the construction group in the Department of Civil Engineering of Hunan University. In 1953, the department was transformed into the Central South Institute of Civil Engineering and Architecture, becoming the leading institute in the Southern Yangzi River (Jiangnan). In 1962, the program was among the first graduate enrollment units of architecture authorized by the State Council. Since passing the professional evaluation for the first time in 1996, the cultivation of undergraduate and postgraduate students has maintained the grade of "excellent" in the many following evaluations. In 2011, the university was granted the right to provide a doctorate degree of the first-class discipline of architecture. The department was approved as a post-doctoral mobile station in architecture in 2014. In 2019, it was approved as a national first-class undergraduate professional construction site.

2. Discipline orientation and features

The degree of Architecture at Hunan Uni versity has four main academic directions: Architectural Design and Theory, Architectural History and Theory, Architectural Technology Science, and Urban Design Theory and Methods. Through scientific research projects and social practice, school has established a serial of featured fields, which include "Local Architectural Creation and Praxis", "Sustainable Architectural Technology", "Green Livable Villages and Towns", and "Digital Protection Technology Of Architectural Heritage".

3. Objectives of professional training

The program of degree strives to inherit the thousand-year history of Yuelu Academy, continue the discipline foundations of the Central South Institute of Civil Engineering and Architecture, follow the comprehensive discipline background of Hunan University, adapt to the trend of globalization and the characteristics of technological change, and strive to cultivate high-level leading talents of architecture for the industry with innovative thinking, high humanistic intuition, solid and broad engineering practice ability.

Introduction to urban and rural planning

1. Discipline overview

The degree program at Hunan University is among the earliest ones in China to provide planning education. It has a complete professional training system, from undergraduate, academic, and professional postgraduate programs to academic and engineering doctoral and postdoctoral programs, and it is considered a double first-class key department in Hunan Province. Both the undergraduate and graduate education tracks have passed professional evaluation and are valid for 6 years.

2. Discipline orientation and features

This degree program includes five academic areas: Urban and Rural Planning and Design, Housing and Community Construction Planning, Urban and Rural Ecological Environment and Infrastructure Planning, Urban and Rural Development History and Heritage Protection Planning, and Regional Development and Spatial Planning. Through scientific research projects and social practice, the program has established a serial of featured fields, and provides curriculums for urban spatial structure, urban public safety and health, hilly urban planning and design, rural planning, urban renewal, and community construction. The program provides access to the Hunan Key Laboratory on the Science of Urban and Rural Human Settlements in Hilly Areas, the Hunan Provincial Land and Space Planning Research Center that was jointly built with Hunan Provincial Department of Natural Resources, and the China Academy of Urban and Rural Construction and Social Governance which was jointly organized with the Ministry of Housing and Urban Rural Development.

3. Objectives of professional training

The program focuses on the cutting-edge theories, tackles major national needs and the problems surrounding individual quality of life, serves national and local economic strategies, undertakes national scientific research tasks, produces high-level academic achievements, provides high-quality planning, design, and consulting services, plays a role in local targeted poverty alleviation and rural revitalization, leads the preparation of local construction standards, and promotes the development of professional academic organizations. We are committed to cultivating high-caliber talents with a solid educational foundation, broad vision, political integrity and talent, high moral compass, innovative thinking abilities, and exploratory spirit.

总体课程概述

陈翚 建筑系主任

我院建筑学专业主干课程组织注重落实《全国高等学校建筑学专业评估文件（2018年版·总第六版）》《普通高等学校本科专业类教学质量国家标准》《湖南大学本科专业培养方案修订意见（湖大教字〔2019〕29号）》等文件的相关要求，主要强调：（1）课程之间的协同与前后衔接，形成特色化课程体系（跨年级）和课程包（同年级为主）；（2）课程本身的前沿性，增加行业发展前沿理论课，选修课强调前沿性与理论化，小课（16课时）为主；（3）人才培养的多元特色，对接本硕博的一贯制培养；（4）跨学科的培养体系，与土木工程学院等单位协同开展新工科建设。

一、课程建设组织

1. 以设计课程为主干的课程包建设

建筑学专业教育，以"设计基础I、II"和"建筑设计I-VI"构成主干课程，围绕主干课程建设，设置"设计基础知识""建筑历史与理论""建筑技术""建筑执业基础""城市设计""计算机模拟与辅助设计"等知识体系。

各年级的课程，以"建筑设计"为核心建设为"课程组"，将各个知识点关联，理论教学与设计实践相配合，形成一个完整的主题化训练单元。各个年级之间，从"设计基础I、II"到"建筑设计I-VI"，再到"毕业设计"，按照时间序列，由浅入深、由简单到复杂、由单体到城市，渐进式开展设计主题训练。通过实施以设计主干课为主线的课程组建设，串联知识模块。其中一年级为建筑基础教育阶段，以"形式与认知"为主题建设统一的学科基础课平台；二到四年级为建筑专业教育阶段，二年级以"空间与环境"、三年级以"建构与营造"、四年级以"技术与综合"为主题分别组织专业核心课与专业选修课；五年级则以毕业设计为核心，围绕"创作与实践"的主题，培养多专业综合协调能力与建筑知识的综合运用能力。

目前，建筑设计主干课程建设已经形成大型木构建筑原型（一年级，单元空间＋景观小品）、微空间（二年级，

中小型建筑：包括建筑群体和单体空间与地域环境、城镇与乡村环境的综合设计训练）、场所文脉（三年级，城市文脉＋城市记忆博物馆）、群组建筑与再利用设计（三年级，群组设计＋教育建筑）、生态高层建筑（四年级，高层建筑＋可持续建筑）、数字大跨（数字建筑＋大跨建筑）等教学主题。

2. 特色前沿课程建设

2008年以来，我院开展了一系列的教学改革研究，提高建筑学本科的教学效果与办学效率。目前，已经在设计基础课程、数字建筑教育、实践课程体系建设、毕业设计的量化评分方法与反馈评价机制等领域形成特色。

二、教学特色梳理

1. 梳理课程体系层级

对应建筑学一级学科专业课程指南，构建"一轴两翼"的专业课程教学组织与层级体系，即以建筑设计课程为主轴，梳理基础理论课程和前沿特色课程的教学目标、内容、形式与时序，形成完整清晰的课程包体系。

2. 重视双语能力培养

提升国际合作与交流的质量，拓展全球学术视野。多年来，聘请海外名师，联合境外一流高校，持续开展"长周期设计课程国际同步教学"（中国与捷克、中国与斯洛文尼亚）"中意俄线路遗产保护""中意数字仿真技术""湖南大学与台湾铭传大学移地教学实践""中日东亚都市比较研究"以及"村镇活化"等教学合作项目，开设"建筑理论""遗产活化"等全英文或双语课程，培养国际前沿资讯、信息的获取能力。

3. 重视专业前沿把控

开设"岳麓建筑讲堂"，聘请海内外学者、新锐建筑师走进课堂，讲授前沿理论、创作方法以及执业方法。

4. 强调实验实践能力培养

依托数字遗产、建筑虚拟仿真、建筑新媒介等实验室建设，开设相关课程，提升数据采集、数据分析、空间模拟能力。

Overall Course Overview

Chen Hui, Director of architecture Department

The organization of the main courses of architecture specialty of our college pays attention to the implementation of the relevant requirements of the national architectural specialty evaluation document of colleges and universities (2018 version · general version 6), the national standard for the teaching quality of undergraduate majors in ordinary colleges and universities, and the revision opinions on the training scheme of undergraduate majors of Hunan University (hudajz [2019] No. 29), It mainly emphasizes: 1) the coordination and connection between courses to form a characteristic curriculum system (cross grade) and curriculum package (mainly in the same grade); 2) The cutting-edge nature of the course itself increases the cutting-edge theoretical courses of industry development, and the elective courses emphasize cutting-edge and theorization, mainly small courses (16 class hours); 3) The diversified characteristics of talent training are connected with the consistent training of this, master and doctor; 4) Interdisciplinary training system, cooperate with civil engineering college and other units to carry out the construction of new engineering subjects.

1. Curriculum construction organization

1.1 Curriculum package construction based on Design Curriculum
The education of architecture specialty consists of "basic design I and II" and "architectural design I-VI". Around the construction of the main curriculum, it sets up knowledge systems such as "basic design knowledge" "architectural history and theory" "architectural technology" "Fundamentals of architectural practice" "urban design", "computer simulation and aided design".

The courses of all grades take "architectural design" as the core and build it into a "course group", which connects various knowledge points and combines theoretical teaching with design practice to form a complete thematic training unit. From "basic design I and II" to "architectural design I-VI" to "graduation design" among all grades, design theme training is carried out gradually from shallow to deep, from simple to complex, from monomer to city according to time series. Through the implementation of curriculum group construction with the main design course as the main line, the knowledge modules are connected in series. The first grade is the stage of architectural basic education, and a unified basic course platform is built with the theme of "form and cognition"; The second and fourth grades are the education stage of architecture specialty. The second grade organizes professional core courses and professional elective courses with the theme of "space and environment", the third grade with the theme of "construction and construction", and the fourth grade with the theme of "technology and integration"; The fifth grade takes the graduation design as the core and focuses on the theme of "creation and practice" to cultivate the comprehensive coordination ability of multiple majors and the comprehensive application ability of architectural knowledge.

At present, the construction of main courses of architectural design has formed large-scale wooden building prototype (grade 1, unit space + landscape sketch), micro space (grade 2, small and medium-sized buildings: including architectural groups and single space and regional environment: including comprehensive design training of urban and rural environment), place context (grade 3, urban context + urban memory Museum), Group building and reuse design (grade 3, group design + educational building), ecological high-rise building (grade 4, high-rise building + sustainable building), digital long-span (digital building + long-span building), etc.

1.2 Characteristic frontier curriculum construction
Since 2008, our college has carried out a series of teaching reform research to improve the teaching effect and school running efficiency of undergraduate architecture. At present, it has formed characteristics in the fields of basic design courses, digital architecture education, practical course system construction, quantitative scoring method and feedback evaluation mechanism of graduation design.

2. Combing teaching characteristics

2.1 Sort out the level of curriculum system
Corresponding to the professional curriculum guide of the first-class discipline of architecture, build a professional curriculum teaching organization and hierarchical system of "one axis and two wings", that is, take the architectural design curriculum as the main axis, sort out the teaching objectives, content, form and sequence of basic theory courses and cutting-edge characteristic courses, and form a complete and clear curriculum package system.

2.2 Pay attention to the cultivation of bilingual ability
Improve the quality of international cooperation and exchanges and expand global academic vision. Over the years, famous overseas teachers have been employed to jointly carry out "international synchronous teaching of long-term design courses" (China and the Czech Republic, China and Slovenia) "Sino Italian Russian line heritage protection" "Sino Italian digital simulation technology" "land transfer teaching practice of Hunan University and Taiwan Mingchuan University" "comparative study of East Asian cities between China and Japan", and "Village activation" and other teaching cooperation projects, set up "architectural theory" "heritage activation" and other all English or bilingual courses, and cultivate the ability to obtain international cutting-edge information and information.

2.3 Pay attention to professional frontier control
Set up "Yuelu Architecture Lecture Hall" and invite scholars and cutting-edge architects at home and abroad to enter the classroom to teach cutting-edge theories, creative methods and practice methods.

2.4 Emphasizing the cultivation of experimental and practical ability
Relying on the laboratory construction of digital heritage, building virtual simulation and building new media, set up relevant courses to improve the ability of data collection, data analysis and spatial simulation.

目录

课题三：单元空间组合设计   42
Topic 3: Unit Space Combination Design

课题四：建筑单体强化设计   48
Topic 4: Strengthening Design of Single Building

三年级建筑设计课程介绍
Course Introduction of Architectural Design Course for Grade Three

课题五：捷克 Liberec 博物馆改扩建设计　　　128
Topic 5: Reconstruction and Expansion Design of Liberec Museum

课题六：汉口胜利仓库改扩建设计　　　136
Topic 6: Reconstruction and Expansion Design of Shengli Warehouse in Hankou

四年级建筑设计课程介绍　　　143
Course Introduction of Architectural Design Course for Grade Four
课题一：高层建筑设计　　　148
Topic 1: High-Rise Building Design

城市设计课程介绍　　　　　　　　　　　　　　　　　　　　　217
Course Introduction of Urban Design
课题一：凤凰县老城区旧城更新城市设计　　　　　　　　　224
Topic 1: Urban Design of Phoenix County Old Town Old Town Updates

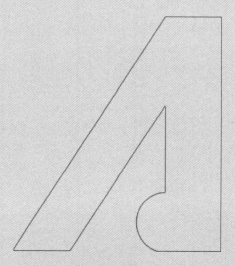

湖南大学建筑与规划学院
School of Architecture and Planning, Hunan University

二年级

Second grade

二年级建筑设计课程介绍
Course Introduction of Architectural Design Course for Grade Two

课程内容：建筑设计 I、建筑设计 II

Course content: Architectural Design I, Architectural Design II

教师团队
Teacher team

叶强
Ye Qiang

谢菲
Xie Fei

杨涛
Yang Tao

苗欣
Miao Xin

向昊
Xiang Hao

李煦
Li Xu

何成
He Cheng

余燚
Yu Yi

李理
Li Li

课程介绍
Course introduction

本课程是建筑学专业本科生的主干专业核心课程。二年级是建筑设计的正式入门阶段，建筑设计 I 主要通过传授建筑设计的基础理论、基本知识，培养学生初步设计能力，围绕建筑设计的空间、环境两个核心设计内涵和外延，分别有针对性地展开有关主题的设计知识的传授和技能训练，建立问题导向、批判性思考的职业习性，从专业特色技能角度培养正确的设计思维方法，树立正确的设计观，为下个阶段的建筑空间与环境综合设计、综合问题解决能力、综合思维等教学目标和知识体系建构奠定基础。掌握单元空间、单功能建筑设计创作的一般规律与方法；对建筑尺度（中观尺度）的尺度感、建筑与环境的关系有一定的认识；掌握简单场地总图设计的一般规律与方法；了解建筑设计与结构选型的基本知识。通过实地调研、课程设计作业，加强制图、表达、表现能力以及对空间与环境的理解、公共建筑设计原理的学习。同时通过增加快速构思的训练，加强学生草图表达能力。

建筑设计 II 以 "空间与环境" 为核心，课程教学重点为中小型建筑（包括建筑群体和单体）空间与地域环境（包括城镇与乡村环境）的综合设计训练，目标为培养学生对空间和环境综合问题的认知和解决能力。课程设计立

This course is the backbone of architectural design for architecture majors. The second year is the introductory stage of architectural design. Design I mainly teaches the fundamantal theory and basic knowledge of architectural design, and cultivates students' basic design ability. Focusing on the two core design connotation and extension of space and environment of architectural design, it carries out design knowledge teaching and skill training on relevant topics respectively, and establishes the professional habit of problems-based critical thinking, cultivates the correct design thinking method from the perspective of professional characteristic skills, and establishes the correct design concept, so as to lay a foundation for the construction of teaching objectives and knowledge system such as comprehensive design of architectural space and environment, comprehensive problem-solving ability and comprehensive thinking in the next stage. Master the general rules and methods of unit space and single-function architectural design; Have a certain understanding of the scale sense of architectural scale (medium architectural scale) and the relationship between architecture and environment; Master the general rules and methods of simple site general plan design; Understand the basic knowledge of architectural design and structural selection. Through research and project design,we strengthen the drawing, expression, expressive ability, understanding of space and environment, and the study of the principles of public building design. At the same time, through the addition of rapid construction, students' sketch expression ability can be strengthened.

Architectural Design II takes "space and environment" as the core. The course teaching focuses on the comprehensive design training

足于具体教学情境的综合视角，确立教学内容。基于湖湘丘陵山地特征和人文习俗，借助地域主义、建构主义、居住建筑设计等现代建筑设计理论，课程教学围绕建筑学科特色，以技能实训和小班教学为主，来培养学生地域建筑的创新设计能力。课程主要分为两个阶段：第一阶段，通过对地域性传统民居聚落空间的研讨和当代群体空间设计原理的传授，让学生在基地、人群调研的基础上，梳理出具有地方性的场地文脉、行为、社区、空间的特征，在有针对性的基地环境特征内外条件的双向制约下，理解和掌握生成总体设计概念的方法以及进行群体建筑形态组织的技巧；第二阶段，在上一阶段设计成果的基础上，对单体建筑的空间形态、功能组织、结构选型和材料等进行深化设计，探究单体建筑空间形态、室内外微环境、功能配置和当地人群行为语境之间的逻辑性，探索具有地域生态性空间的生成方式及地域材料的创造性运用。

整个二年级的课程题目涉及不同的建筑类型，让学生自由发散地进行选题，训练学生对空间形态（室内外）的感知和设计能力，并使其树立建筑与环境相互结合的观念，通过多元多样的选题，既提升学生对建筑设计的兴趣，又能让其在设计创作的过程中找到乐趣，为此赋予情趣，这也是二年级课程设计的宗旨。

of the space and regional environment (including urban and rural environments) of small and medium-sized buildings (including building groups and individual buildings). The goal is to cultivate students' cognition and ability to solve comprehensive problems of space and environment. Curriculum design is based on the comprehensive perspective of specific teaching situations and establishes the teaching content. Based on the characteristics of the hills and mountains in Hunan and humanistic customs, with the help of modern architectural design theories such as regionalism, constructivism, and residential architectural design, the curriculum teaching focuses on the characteristics of the architectural discipline, skill training and small class teaching to cultivate students' regional architectural innovation design ability. The course is mainly divided into two stages: the first stage, through the discussion of the regional traditional residential settlement space and the teaching of contemporary group space design principles, so that students can sort out the characteristics of local history, behaviors and community on the basis of base and crowd research. Under the two-way restriction of the internal and external conditions of the targeted base environment characteristics, understand and master the methods of generating the overall design concept and the skills of group building form organization; In the second stage, on the basis of the design results of the previous stage, we deepen the design of the space form, functional organization, structure selection and materials of the single building, and explore the space form of the single building, indoor and outdoor micro environment, functional configuration and the logic between the behavioral contexts of local people,and explore the generation of regional and ecological spaces and the creative use of regional materials.

The topics of the entire second-year course involve different types of buildings, allowing students to choose topics freely and divergently, training students to perceive and design the spatial form (indoor and outdoor), and to establish the concept of the integration of architecture and environment. The diversified selection of topics not only enhances students' interest in architectural design, but also allows them to find fun in the process of design creation. For this reason, it is the purpose of the second-year curriculum design.

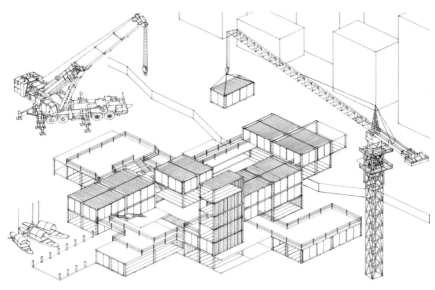

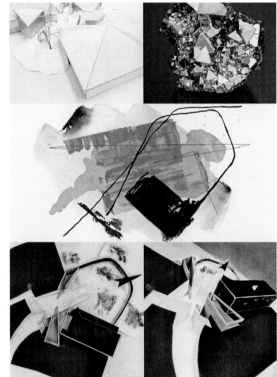

目前，我院二年级建筑课程设计选题主要包括以下几个方向：

1. 发散创作和功能赋予练习，目的是培养学生对空间的感知能力和对平面图形三维化的造型能力。

2. 单元空间设计创作，目的是培养学生对日常生活和学习空间的认知及手绘创作表达能力。

3. 单元组合空间设计，目的是培养学生对不同单元空间的组合能力，尝试处理包括功能布局、场地设计、环境要素以及流线交通在内的复杂问题。

4. 建筑单体强化设计，目的是指导学生解决复杂的空间和形体关系，拓展知识面，完善和塑造更加丰富的建筑空间。

At present, the topics for the second-year architecture course design of our college mainly include the following directions:

1. Divergent creation and function assignment exercises, which aims to cultivate students' perception of space and the ability of three dimensional modeling of plane graphics.

2. Unit space design and creation, which aims to cultivate students' cognition of familiar daily life and learn space and the ability of hand painted creation and expression.

3. Unit combination space design, which aims to cultivate students' ability to combine different unit spaces and try to deal with complex issues including functional layout, site design, environmental elements, and streamlined traffic.

4. Strengthening the design of the building unit, which aims to guide students to solve the complex space and physical relationship, expand their knowledge, improve and shape a richer architectural space.

发散创作和功能赋予练习

Topic 1: Divergent Creation and Functioning Exercises

一、主题

本课题以"发散创作与功能赋予"为题，旨在考查学生通过发散创作和功能赋予的方式，将对空间想象和构思实现从无到图纸的过程，并在此过程中考验学生通过二维图形提取三维空间的创作能力；在提升学生对空间认知能力的同时，培养学生对创作的兴趣，激发其创作欲，并通过此练习让学生熟悉一套设计创作的方法。

二、设计内容及要求

1. 设计内容

本设计采取自由选图和自行拟定场地的方式，范围不限，可选择位置也不设限制。

2. 设计步骤

（1）以"自由的眼睛"（自我挖掘）自由地选择一个自己认为有趣（或喜欢）的图像。

（2）以所选取的图像为底图，在上面提取出关键的线。

（3）沿着所提取的关键线，利用纸板沿竖向在三维方向立起，进而形成立体的空间状态，表达出所选图像的空间特征。

3. 设计要求

（1）将所选择二维图片转化为与其特征形态相匹配、功能要求相对应的三维空间，并将其通过理性分析置入合理的场地内。

（2）在合理生产三维空间的基础上赋予其功能并进行优化，鼓励空间创新、功能创新、形式创新，体现出空间转化的合理性和趣味性，并表达出建筑的质感和体量感。

（3）图纸表达规范。图纸能够充分表达出作品创作的意图，并且包含设计概念描述和设计说明，比例不限，但需合理考虑尺度和比例。

（4）要求制作精细手工模型，鼓励手绘图纸表达。

作品来源：黄心怡

课题二
单元空间设计创作
Topic 2: Design and Creation of Unit Space

一、主题

本课题以"单元空间设计创作"为题，旨在考查学生对独立空间单元的塑造能力和打磨能力，以学生身边日常使用的空间为创作对象，通过观察、实地调研、寻找参考案例、尺度测绘、概念构思等环节，聚焦小范围、小尺度的空间设计创作，带有批判性和开放性的思考方式，寻找以学习空间为代表的空间需求的本质，思考人与空间的关系。

二、设计内容及要求

1. 设计内容

本设计采取拟定目标对象和拟想目标对象的方式，设计对象为和学生日常生活相关的单元空间。

2. 设计步骤

（1）选定身边常用的、熟悉的、喜爱的单元空间。

（2）在对选定对象进行优劣评判和对比分析的基础上，对相关空间进行详细测绘。

（3）结合优秀案例分析和实地调研，有选择性地进行方案创作，创作思路包括局部改造、理想空间重塑、特色空间构想等。

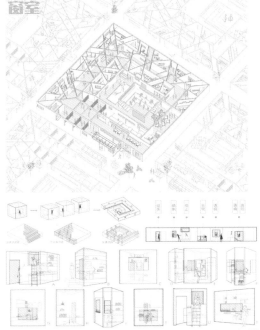

作品来源：牛冰倩

作品来源：梁代君

3. 设计要求

（1）完成可容纳人数与空间尺度相匹配的拟定空间整体方案设计。

（2）充分考虑使用者的需求，鼓励空间形态创新、功能创新、设计方法创新。

（3）图纸表达规范准确。图纸能够充分表达出作品创作的意图，效果图清晰展示空间特征。

（4）要求制作精细手工模型。

课题三

单元组合空间设计
Topic 3: Unit Space Combination Design

一、主题

本课题以"单元组合空间设计"为题，指定校园周边具有代表性的校园生活园区、工业园区、艺术园区等为调研对象和选址范围，基于对园区内各空间的了解和分析，在既有园区的规划用地范围内，融入校园文化、市井文化、办公文化等其他诸多特色要素，发挥设计者的空间想象能力和空间塑造能力，打造包含各类文化元素在内的多功能复合空间。旨在考查学生的场地意识和不同空间类型的组合能力。

二、设计内容及要求

1. 设计内容

本设计采取指定范围选址的方式，要求设计者充分理解指定场所的特征，通过对题目阅读，挖掘该场地所传达出的文化内涵和文化特质。在解读校园文化和艺术园区关系的基础上合理地融入其他文化形式，尝试策划城市空间事件（如创意活动、公共艺术、电子媒体技术等），鼓励实验性、创新性、探索性地挖掘多元文化结合实践的可能性，在校园环境背景下塑造有价值、有意义、有内涵的空间形态；让设计者所营造的复合空间成为讨论、交流、培育城市文化事件的发生器，成为有时代人文气息、有邻里温度的空间活动场所。

2. 设计要求

（1）空间的功能定义需结合设计者的策略和概念来进行整体策划和统筹，合理体现对已有文化的传承和对融入文化的接纳，要满足通过设计来改善和丰富现有校园文化生活的初衷。

（2）通过调研和分析，基于设计策略和构思制定设计任务书技术指标（包括用地面积、占地面积、总建筑面积、容积率等），并完善和整合设计概念。

（3）复合空间的出入口位置需要根据调研和设计策略合理规划设置，要注意复合空间的建筑形态以及各单元空间之间的衔接。

（4）模型必须能够清晰地看到内部的陈设和布局。

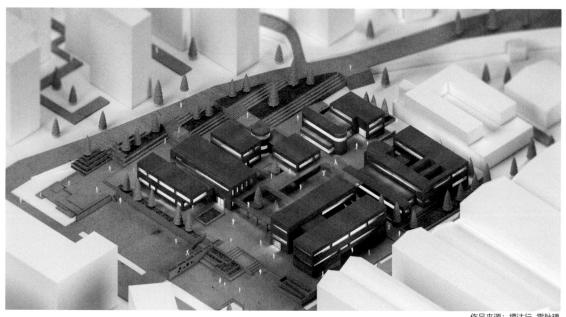

作品来源：禤法行 雷秋瑀

课题四
建筑单体强化设计
Topic 4: Strengthening Design of Single Building

一、主题

本课题以"建筑单体创作"为题,在单元组合空间和场地设计的基础上优化单体设计,旨在考查学生对建筑空间的细化能力,需要精准把控尺度比例、材料节点、交通流线、功能分区、竖向设计和立面造型等设计准则,考虑如何以设计思维和技术手段去解决现实社会的问题,将场地设计和建筑设计融合在一起。

二、设计内容及要求

1.设计内容

本设计在延续单元空间组合的基础上,针对性地细化建筑单体空间。

2.设计要求

(1)建筑单体需要与单元组合空间之间相互呼应,有起承和衔接的关系。

(2)充分融入场地设计,细化图纸深度,总平面和各层平面的绘制达到技术图纸绘制要求。

(3)图纸能够充分表达出作品创作的意图,效果图清晰展示空间特征。

(4)要求制作精细手工模型。

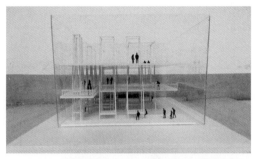

作品来源:李正刚

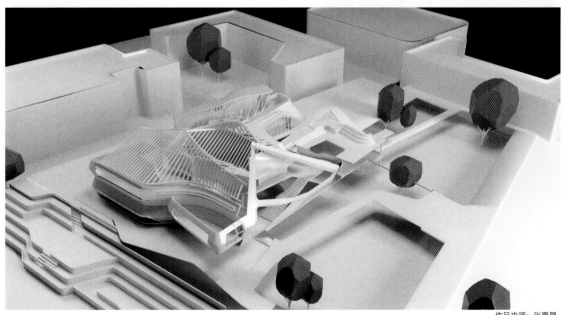

作品来源:张嘉晟

设计题目：设计初探 X——"窥探边界"后湖游船码头空间设计
指导老师：李理
学　　生：陈俊海

1

Design topic: Preliminary Design X— "Spying the Boundary" Houhu Cruise Ship Wharf Space Design

Instructor: Li Li

Student: Chen Junhai

● **设计说明**

后湖是一片开阔的人工湖，中间有小沙洲，站在其中，远处的岳麓群峰清晰可见。南边是正在建设的商业区和居住区。原船坞地处隐蔽，入口狭小，有 6~9 艘船。陆与湖、水面上下、船与水与人、景物、人造自然和城市……。无处不在的边界提醒我们要在城市中重新思考人与自然的关系。

学习可能对应身体的不同状态和不同姿势，尽可能地感受利于学习的氛围，同时又要尽可能不被非相关因素影响。这种倾向，我认为是对内的活跃与对外的隔离。在寻求智慧的过程中，最理想的状态应当是与世隔绝但并不自我封闭，从而纯粹地进行思考并获得提升。使用在弧面上不断变化的光影和空气的流动，通过封闭、开放、隔断的安排，促成这种理想状态的形成。当完成这种"似是而非的封闭与开放"的状态时，思想的纯粹性也应该自然而然地构建出来。

Design notes

Houhu is an open artificial lake with a small sandbar in the middle. Standing on it, we can see the peaks of Yuelu Mountain in the distance clearly.. There are commercial areas as well as the residential areas under construction. The original dock is a covet place and the entrance is small, with 6-9 ships. Land and lake, above and below water, boat and water and people, scenery, man-made nature and city... The ubiquitous boundary reminds us to rethink the relationship between man and nature in the city.

Learning may correspond to different states and different postures of the body, we should feel the atmosphere conducive to learning as much as possible and at the same time not to be affected by unrelated factors as far as possible. This tendency, I think, is the internal activity and external isolation. In the process of seeking wisdom, the ideal state should be to be isolated from the world but not to be self-enclosed, so as to purely think and get promoted. Using the constantly changing light and shadow and air flow on the curved surface, through the arrangement of closed, open, and partition, we promote the formation of this ideal state. When this "paradoxical closed and open" state is completed, the purity of thought should also be constructed naturally.

边界：湖岸　　边界：水舟　　边界：虚实　　边界：荷露

设计题目: 点子·手段·空间生成/观念·功能·场所营造
　　　　　——"轮转"后湖大学生文娱场所
指导老师: 李理
学　　生: 唐艾乐

2

Design topic: Ideas · Means · Space Generation / Concept · Function · Place Creation
　　　　　　—"Rotating" Houhu University Students Recreation Venue
Instructor: Li Li
Student: Tang Aile

● **设计说明**

通过对基本事物图片的线条提取和深化获取建筑的基本形态,再对有形的建筑赋予功能,以此熟悉从形式到功能再到建筑设计的一种基本方法。本方案则是通过对雨花石图片的基本找形进行的展开,最终设计出一个位于湖边小岛上的小型建筑群,主要为大学生提供娱乐和学习的功能。

Design notes

By extracting and deepening the lines of the pictures of basic things, we can obtain the basic form of the building, and then endow the tangible building with functions, so as to be familiar with a basic method from form to function and then to architectural design. This scheme is based on the basic shape finding of the Yuhua stone picture, and finally designs a group of small buildings located on the lake island, which mainly provides entertainment and learning functions for college students.

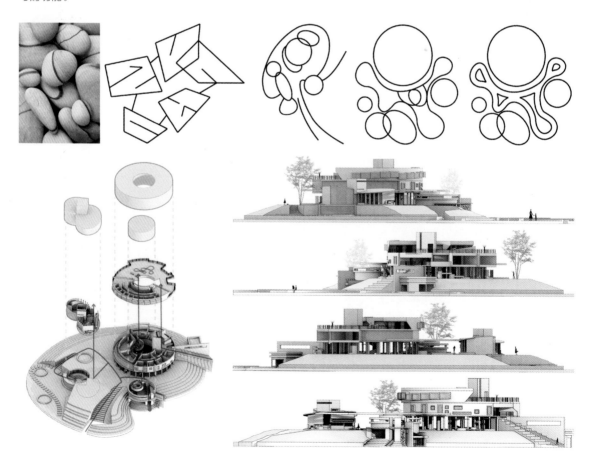

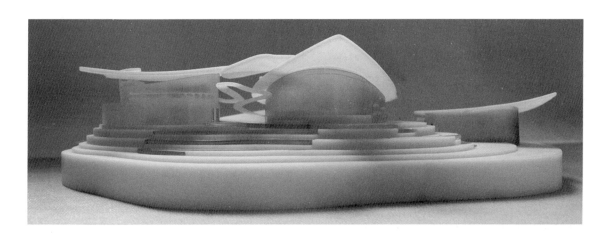

14

设计题目： 点子 · 手段 · 空间生成＆观念 · 功能 · 场所营造
　　　　　　—— 后湖儿童活动空间设计
指导老师： 李理
学　　生： 刘加正

3

Design topic: Idea · Means · Space Generation & Concept · Function · Place Creation
　　　　　—Houhu Children's Activity Space Design
Instructor: Li Li
Student: Liu Jiazheng

● 设计说明

此设计位于后湖的湖中小岛上，灵感来源于树枝上的几片新叶。对叶子的线条进行提取后，经过一步步空间操作，最终形成了现在的建筑形态。后将游乐、餐饮、阅读、VR 体验等功能赋予空间，推敲出了该设计的室内布置。小建筑群犹如几片轻盈又生机盎然的新叶，生长在环境优美的湖心小岛上，为小朋友们提供了一个快乐的活动场所。

Design notes

This design is located on a small island in the middle of Houhu lake and is inspired by several new leaves on the branches. After extracting the lines of leaves, the current architectural form is finally formed through step-by-step spatial operation. After that, the functions (entertainment, catering, reading, VR experience)are given to the space, and the indoor layout of the design is deduced. The small building complex is like several light and vibrant new leaves, which grow on the small island in the middle of the lake with beautiful environment, providing a happy place for children.

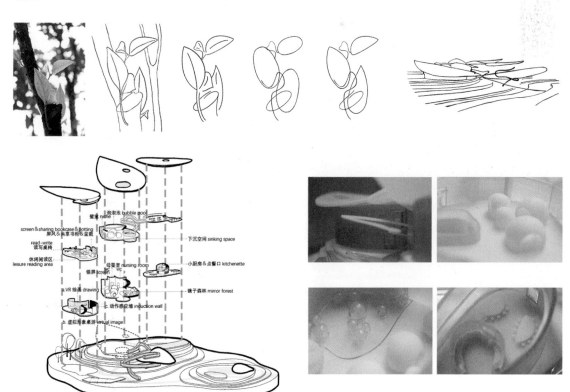

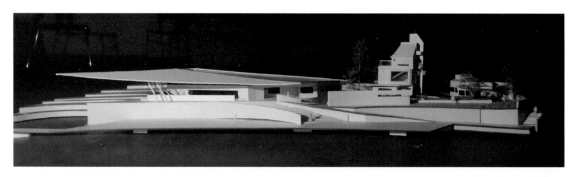

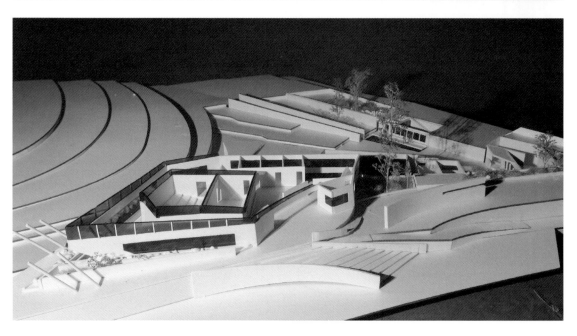

4

设计题目： 点子·手段·空间生成／观念·功能·场所营造
　　　　　 ——"展望"后湖大学生科研基地设计
指导老师： 李理
学　　生： 姜佳辰

Design topic: Idea · Means · Space Generation & Concept · Function · Place Creation
　　　　　 —"Prospect" Houhu University Student Research Base Design

Instructor: Li Li

Student: Jiang Jiachen

● 设计说明

本设计的概念受原始图片的启发，由具有延展性的线条联想到对未来世界及未知领域的展望。后湖作为一个文创基地，未免有些太过文艺，缺乏一些新锐而偏向实际领域的研究，与此同时，靠近天马学生宿舍的后湖北岸恰有一块三角形的荒地，同湖对岸的城市遥遥相望。由于其视野开阔，且地块形状同方案第一阶段的平面十分吻合，故选用此地块作为继续深化方案的基地。"展望"一概念在本设计中主要体现有三：一为形式，延展外放；二为功能，天文基地是其一，科研探索同样是一种展望；三曰精神，人类的探索精神永世长存，引领我们走向远方。

Design notes

Inspired by the original picture, the concept of this design is associated with the prospect of the future world and unknown fields from the malleable lines. As a cultural and creative base, Houhu is too literary and artistic, lacking in creative and practical research. At the same time, close to Tianma students' dormitory reserved a triangular land on the North Bank of the lake, facing the city on the other side of the lake. Due to its wide field of vision and very consistence with the scheme in shape, this plot is selected as the base for deepening the scheme further. The concept of "envisage" is mainly embodied in this design in three aspects: Firstly in form, extension and outing; Secondly in function, the astronomical base is one of them, and the scientific exploration is also a kind of prospect; Thirdly is spirit, our instinct of exploration will last forever and lead us to the distance.

● 生成过程

Generation progress

1. 原始图片搜寻　　**2. 特征线提取**　　**3. 平面草图**　　**4. 体块生成**　　**5. 深化空间／第一阶段成果**

 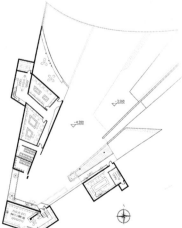

设计题目： 瑜伽会馆设计 —— 类型人群行为空间
指导老师： 苗欣、邓广
学　　生： 李镜宇

5

Design topic: Yoga Club Design—Type Crowd Behavior Space

Instructors: Miao Xin，Deng Guang

Student: Li Jingyu

● **设计说明**
Design notes

此建筑选用轻钢结构，可以减轻建筑自重，减小墙壁厚度，使建筑视觉效果轻盈。通过光照分析可以发现，此建筑清晨直接受光面积不大，但由于墙体少，漫反射可使室内光线充足。清晨，场地较为安静，也是最适合练习冥想的时间，冥想室室内的墙壁上刚好有一道光带，易于营造较好的环境。下午日照较为强烈，不适合练习瑜伽，可加遮挡物或做其他用途。

This building adopts light steel structure, which can reduce the weight of the building and the thickness of the wall and make the visual effect of the building light. Through the light analysis, it can be found that the direct light receiving area of this building in the morning is small, but due to the few walls, the diffuse reflection can make the light sufficient indoor. In the early morning, the buildings are relatively quiet, which is also the most suitable time to practice meditation. There is just a light band on the wall of the meditation room, which is easy to create a better environment. The sunshine is strong in the afternoon, so it is not suitable for practicing yoga but shelter can be added or apply it for other purposes.

● **行为分析**
Behavior analysis

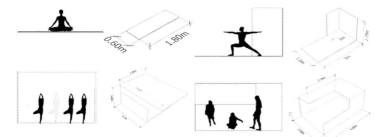  

● **功能分区**
Functional division

 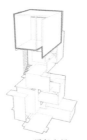   

练习空间　　　　冥想空间　　　　休息空间　　　　交通空间　　　　灰空间
practice space　　meditation space　　relax space　　transport space　　gray space

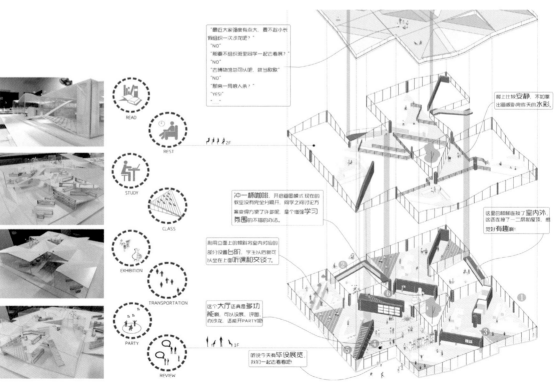

设计题目： 斜·折 | 6000m³ 体积限定下的建筑训练营设计
指导老师： 李煦
学　　生： 张可心

6

Design topic: Oblique · Folding | Architectural Training Camp Design under the Limit of 6000m³ Volume

Instructor: Li Xu

Student: Zhang Kexin

● 任务书要求

在 划 定 的 6000m³ 的 体 积 内 设 计 出 一 个 4500m³(±20%) 的 有 效（封 闭）的 建 筑 空 间 与 1500m³(±20%) 的 外 部 空 间。4500m³(±20%) 的 建 筑 空 间 必 须 容 纳 ≥9 个 基 本 的 学 习 单 元 空 间（学 习 单 元 空 间 体 积 自 定）及 相 应 辅 助 空 间，并 且 在 空 间 之 间 形 成 有 效 而 且 有 趣 的 组 织 关 系。空 间 的 功 能 定 义 为 建 筑 训 练 营 地，空 间 的 使 用 者 包 括 但 不 局 限 于 建 筑 学 和 相 关 专 业 的 学 生 和 老 师，每 个 学 习 单 元 必 须 容 纳 30 人。设 计 构 思 必 须 找 到 一 个 支 撑 点 作 为 起 始（如 一 幅 画、一 个 故 事、一 部 电 影、格 网、肌 理、环 境 景 观 元 素）。

Mandate requirements

Design an effective (closed) building space of 4500m³ (±20%) and an external space of 1500m³ (±20%) within the delineated volume of 6000m³. The building space of 4500m³ (±20%) must accommodate ≥ 9 basic learning unit spaces (the volume of the learning unit space is self-defined) and the corresponding auxiliary space, which forms an effective and interesting organizational relationship between the spaces. The function of the space is defined as an architectural training camp. The users of the space are not limited to students and teachers of architecture and related majors. Each learning unit must accommodate 30 people. The design concept must find a support point as a starting point (such as a painting, a story, a film, grid, texture, environmental landscape elements).

● 设计思路　　Design ideas

传统教室　　　　　　未经定义的开放空间　　　　　斜向墙体定义　　　　　　屋顶和地板形态对应

● 方案演进　　Scheme evolution

各单体之间由一条路径相连　　探索不同单体的效果　　根据任务书的体积要求减少　　两种单体组合得到最佳组合方式
　　　　　　　　　　　　　　　　　　　　　　体块数量并优化室内外关系　　同时立面上注重上下的结合与联系

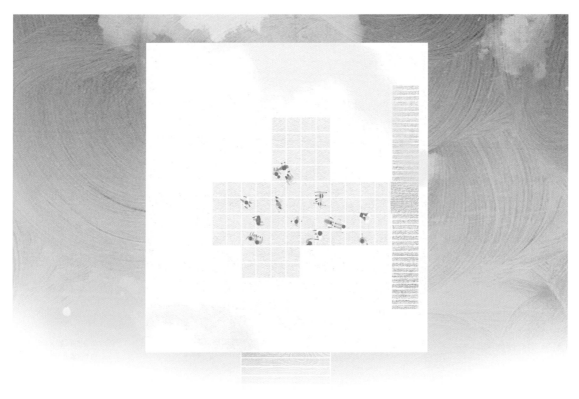

7

设计题目： 海市蜃楼
指导老师： 杨涛
学　　生： 邱子倍

Design topic: Mirage

Instructor: Yang Tao

Student: Qiu Zibei

● **原型变换**　　　　**Prototype transformation**

结构并不只是一门技术，尤其在当代学科分离的情况下，建筑师更应该对结构进行深入思考。从拱出发，来引出建筑系学生对于结构的深思。建筑整体呈下重上轻、下疏上密，似是海中的一座岛屿，而通体的白色给人神圣与不真切之感，在阳光的映照下忽隐忽现，像是海市蜃楼。

拱解放了空间，使之自由开放模数化，并使之遵循秩序，形式得到统一；大面积采光符合建筑学生要求。

The space liberated, the arch, makes it open and follows the order. Its form given large-area daylighting meets the requirements of architecture students.

Structure is not just a technology, especially when contemporary disciplines are separated from each other, architects should think deeply about structure. Starting from the arch, the architecture students are elicited to think deeply about structure. The whole building is heavy and sparse at the bottom, while light and dense at the top. It seems to be an island in the sea, and the whiteness of the whole body gives people a sense of holiness and unreality. The image of the arch, which is like mirage, drifts in and out in the sunlight.

原型变换

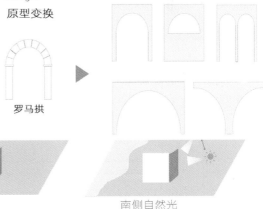

罗马拱

位于中国南海岸

以 19m × 19m 划定平面

抬高 18m

联系岸边与建筑

南侧自然光

夏季东南风，冬季西北风

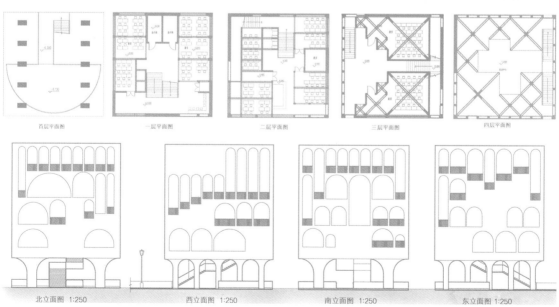

首层平面图　　　　　　　一层平面图　　　　　　　二层平面图　　　　　　　三层平面图　　　　　　　四层平面图

北立面图 1:250　　　　　西立面图 1:250　　　　　南立面图 1:250　　　　　东立面图 1:250

▼ 轴测图

首层与四层相对自由开放，
中间三层主要为教学空间。
各层以二为模数布置阵列，
且原型不同，给人以多样
而又不失和谐的空间感受。

休息　　展览　　讨论

授课　　绘图

授课　　绘图

授课　　绘图

饮食　　演奏　　多媒体

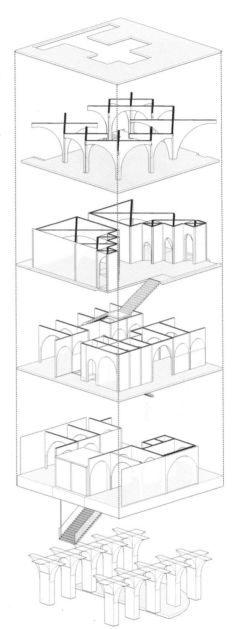

25

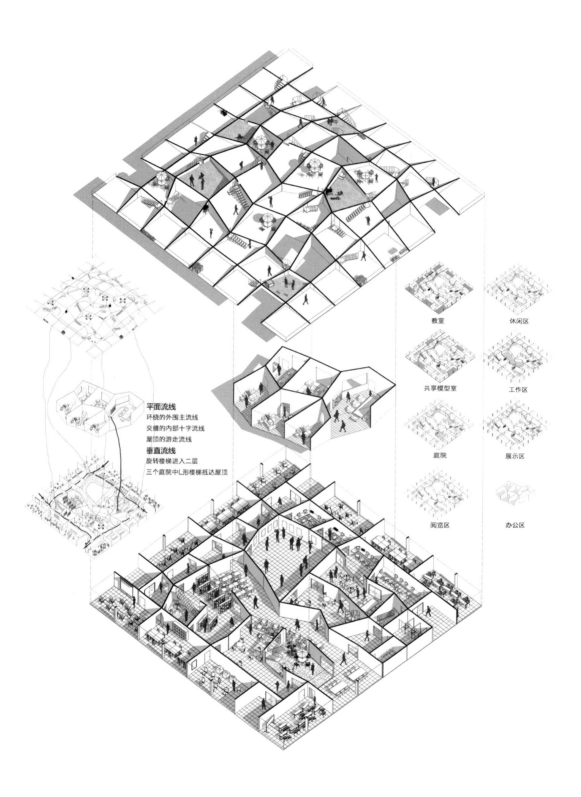

平面流线
环绕的外围主流线
交缠的内部十字流线
屋顶的游走流线

垂直流线
旋转楼梯进入二层
三个庭院中L形楼梯抵达屋顶

教室　　　　　　休闲区

共享模型室　　　工作区

庭院　　　　　　展示区

阅览区　　　　　办公区

8

设计题目：发酵
指导老师：李煦
学　　生：徐正文

Design topic: Fermentation

Instructor: Li Xu

Student: Xu Zhengwen

● **设计说明**

面包中由于酵母的存在，实体被消耗，产生气体，原本的实体被虚体取代，同时膨胀、挤压，产生了一个个孔洞与缝隙，而面包蓬松的口感也正是产生于这实与虚的交错变化中。空间亦然，实与虚永远是辩证存在的，有实才有虚，无虚不谈实。借用面包发酵中的虚实变化生成的法则，能够创造一个丰富多样的空间体，本设计即从此入手。

Design notes

Due to the yeast in the bread, the entity is consumed and gas is produced. The original entity is replaced by the virtual body. At the same time, it expands and squeezes, creating holes and gaps. The fluffy taste of bread is also the result of this staggered and changing reality and virtuality. The same is true for space. Reality and emptiness always exist dialectically. Only when there is reality, there is emptiness. Borrowing the law of the change of reality and emptiness in the fermentation of bread, it can create a rich and diverse space body. This design is based on it.

立方像素

纵横框架

相融相交

加入楼板

灯光场地——发光面砖与像素灯面

要素语言的融合

设计题目： "日常生活"范式的空间建构：望衡对宇 —— 校园服务空间设计

指导老师： 苗欣

学　　生： 韩笑、武文忻

9

Design topic: Construction of "Daily Life" Paradigm: Live Near Each Other—Campus Service Space Design

Instructor: Miao Xin

Students: Han Xiao, Wu Wenxin

● **设计说明**

通过跟踪人群，分别研究青年辅导员和学生创业者，一静一动，如何在 J 字形的地块里融合？因为人群需求的不同，对建筑的要求也不同，因而形成了"若市门庭"与"清幽雅境"这两个看似截然不同的建筑氛围。

但是仔细研究发现，两栋建筑有结合的可能性。而这种结合并不是实体空间的简单相连，而是一种视线上的相交与融合。"望衡对宇"就是对这种关系最直接的表达。所谓"望衡对宇"便是屋檐相邻，相望而不相扰，相对却又欢情相接。激情的舞台如同对话的圆桌，联系着两边的人群；彼时，我们还未相熟；此时，我们为你欢呼。每一次与陌生人的相逢，可能便如此，此时相欢，彼时相散。

Design notes

By tracking the crowd, studying the young counselors and student entrepreneurs separately, how do they blend in the J-shaped plot with each movement? Because of the different needs of the people, the requirements for the building are also different, so two seemingly different architectural atmospheres of "Ruoshimenting"(means boisterous and lively) and "Quiet and Elegance" are formed. However, careful study reveals that there is a possibility of combining the two buildings. And this combination is not a simple connection of physical space, but a kind of intersection and fusion of the line of sight. "Wanghengduiyu" is the most direct expression of this relationship. The so-called "Wanghengduiyu" is the connected eaves , facing each other without disturbing each other, but connected happily. The stage of passion is like a round table of dialogue, connecting the people on both sides; At that time, we were not familiar with each other; while now, we are cheering for you. Every time I meet with a stranger, it may be like this, happy at this time, and separated at that time.

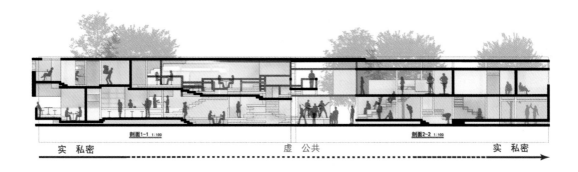

● 调研分析   Investigation and analysis

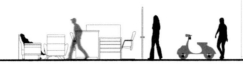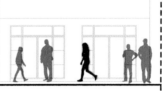

交流时提问者被安排在靠外的沙发上，视线被挡板阻挡，不会被直接看到

会在心理辅导室休息
其休息室隐蔽，不易被发现

电动车出行
会上三把锁

健完身不在健身房洗澡，不愿被认识的人注意，会快速逃离繁忙的街道

研究对象：建院某辅导员
身份：学生、老师

私密性   隐私性

安全性   **高效性**

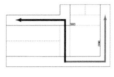
当图示信号灯是红灯时，走红色路线，反之走灰色路线

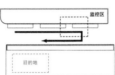
在休闲时，私密性与高效性冲突，会先满足私密性的需求；同时，在某种程度上，高效性在为私密性服务

建筑功能

餐饮区　空间层次丰富的咖啡厅
交流社区　便于交流的吧台
健身区　个人工作室
晾晒区　较为隐蔽的小天台
洗浴区
健身房

服务人群

教职工　　学生　　游客

专注性　　　高效性
半开放性

转换   半私密性

研究对象：建院某学生
身份：草木人禾经营者、吉他爱好者

个人领域
小桌子和椅子高大的靠背围合出一个相对封闭的区域

天台具有开放的属性，但同时又不会被外界观察到，这一半私密性的区域为练琴和聚会提供了合适的场所

在制作咖啡时喜欢与有相同爱好的人交流

步行速度非常快

进行个人工作时非常专注

与同学组有乐队，空闲时会上天台练琴

空闲时会和要好的朋友一起到天台喝酒聊天

30

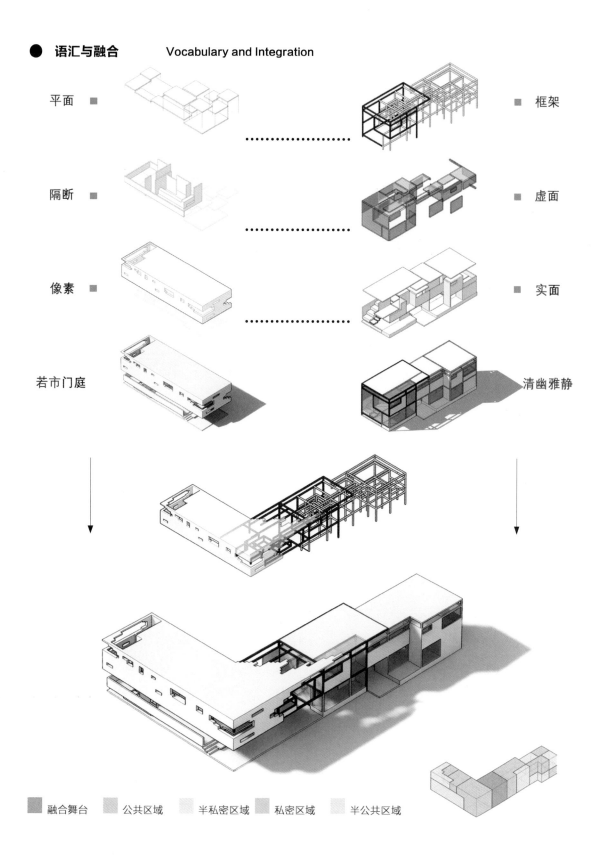

● **语汇与融合**   Vocabulary and Integration

平面 ■

■ 框架

隔断 ■

■ 虚面

像素 ■

■ 实面

若市门庭

清幽雅静

■ 融合舞台   ■ 公共区域   ■ 半私密区域   ■ 私密区域   ■ 半公共区域

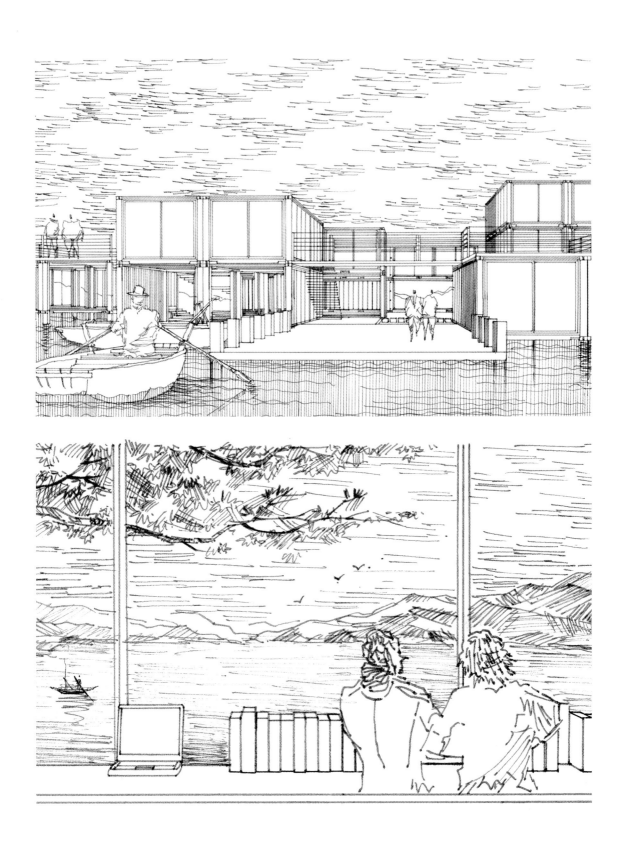

设计题目： 城市漂浮岛 —— 基于学习共同体理论模式的探究
指导老师： 叶强 、胡赞英
学　　生： 和译

Design topic: Urban Floating Island—An Exploration Based on the Theoretical Model of Learning Community

Instructors: Ye Qiang，Hu Zanying

Student: He Yi

● 教学理念沿革
Evolution of teaching ideas

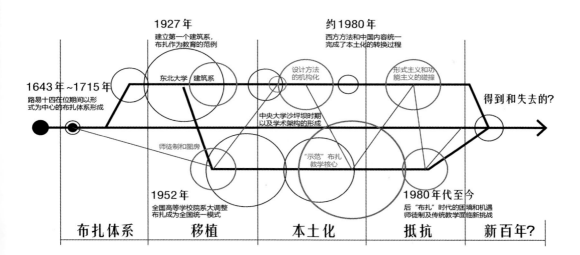

评论类型	简短定义	映射到传统的学术实践
桌边点评	与导师一起、高度生成性、随机学习的认知行为潜能	长度不等、一次强烈的个人投入、学生和教师之间社会互动模式的批判性对话，双方共同解决问题
同伴点评	非正式的、生成性的、自愿的或要求的	利用评论形成深入专业理解基础、扩展评论技能和对设计理念或设计抉择的合理性、逻辑性进行判断
小组评论	半公开的、不太正式、更多生成性的，偶尔评价性	把概念形成和阶段性设计成果作为争论的主要内容、激烈的、效率时高时低
公开评论	公开的、邀请的、总结性的、评价性事件	总结性评论耗时最长，但在观念形成和学习方面效果最弱，陷入说教，具有剧院和表演的某些特征

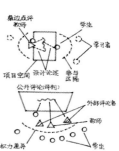

现有理论研究
Existing theoretical research

"学习共同体"的内涵是什么？它仅是"组织和社区"新的概念包装吗？
"共同体"为"学习"提供某种场所、某种语境，或者是某种蕴含时间和历史意义的时空边界，它表示"学习者"所处的特定社会关系或结构。

知识和学习为何要依托"共同体"？
将学习置于共同体语境中加以考察的理论前提是，知识是社会性的建构，学习是意义的协商，而共同体是意义的前提和载体，它具有社会学意义。

如何构建"学习共同体"？
学习共同体的描述性模型包括三个相互作用的特征范畴：情境、指令、催化剂。共同体是众多因素和变量的复合体。互动是针对共同体一致认为重要的东西展开的，这一特征也被称为"对象中心社交"。促进共同体中的沟通需要足够的结构，但是不应令共同体成员受到该结构的束缚。

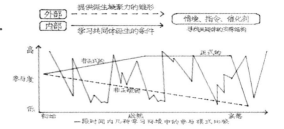

建筑集装箱的引入

集装箱具有模块化、经济集约、方便移动、可回收的特点，它们在内部形成空间并通过钢骨架支撑整体结构，经过改造后能承受最恶劣的天气条件。我们希望设计一个能快速建造并适应环境的漂浮岛，在建筑教育革新的下一个百年中，带着学习乌托邦的理想漂流到世界各地。

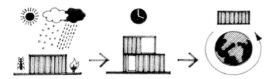

划分出更小的模块

选取标准集装箱作为一个模块 6.3m x 3.3m x 3.3m

模块组成单元空间

四个模块组成一个标准学习单元12.6m x 6.6m x 3.3m

单元空间组成聚落

集装箱组合的基本方式

两个尺寸相同的集装箱，在相同方向上的组合，X、Y和Z轴各有四种对齐关系，产生64种组合结果。其中有8组组合会因为重叠而无效

组合	平行、串联、叠加、错位对角、相交、直立、站立
空间	包容、融合、附着、平台、覆盖、夹心
结构	底座、框架、柱子、墙体、屋面、楼梯、走廊、平台

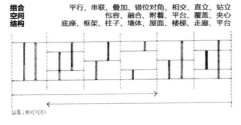

仰览（折可可见）

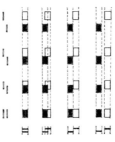

如何让冰冷而笨重的方块在滨水环境中和谐共存，打造一个适宜学习的场所？格式化的呆板样貌，功能至上的框架、不经修饰的外表，将54个格式化单元组合，重构出工业背景下的传统园林意象。建筑与外景相容又自成格局，使室内所有行为都与景随形，与自然相融。

二层及以上底层架空，通过大面积玻璃窗营造建筑的漂浮感，通过码头来增加一个有趣的入口。

与远山和水岸相融：削减体量

漂浮闭环生态系统：自给自足

建筑围合形成封闭庭院，通过人的行走与内向水和外向水发生联系，与喧嚣隔绝，与环境相融。

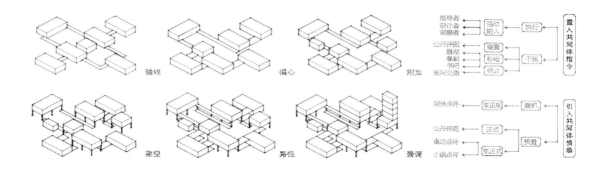

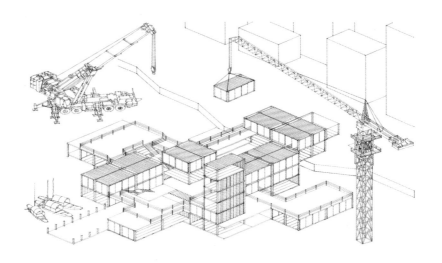

● **室内透视图**
Indoor perspective

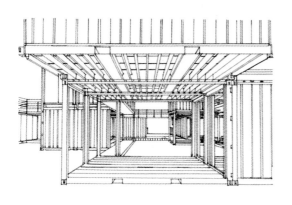 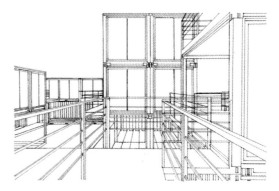

设计题目： Big Toposcape | 校园山地工作室设计
指导老师： 李煦、何成
学　　生： 孙智霖

Design topic: Big Toposcape | Campus Mountain Studio Design

Instructors: Li Xu, He Cheng

Student: Sun Zhilin

● **设计说明**

在国内，真正意义上"没有围墙"的大学并不常见，诸多高校以安全为由围墙高筑，与城市和城市空间切断了联系。湖南大学一直宣称自己是"没有围墙的大学"，但面对校园边界的消极态度与冰冷的围墙无异。设计通过对挡土墙要素在场地中的消极意向（分割）进行反转。在中国南方地区私家园林的典型案例中提取原型并进行重构，使"墙"这一元素成为引导人前行的线索，重新建构位于校园边界的建筑与周围环境的联系，使建筑成为连接自然环境与校园环境的媒介。

Design notes

In China, universities without bounding walls are not really common. Many universities have built high walls on the grounds of safety, cutting off the connection with the city and urban space. Hunan University claims to be a "university without bounding walls", but its negative attitude towards campus boundaries is no different from a cold wall. The design reverses the negative intention (segmentation) of the retaining wall elements in the site. Extracting prototypes from typical cases of private gardens in southern China and reconstructing them, the "wall" element is made as a clue to guide people forward, reconstructing the connection between the building on the campus boundary and the surrounding environment, making the building a medium to connect nature and campus environment.

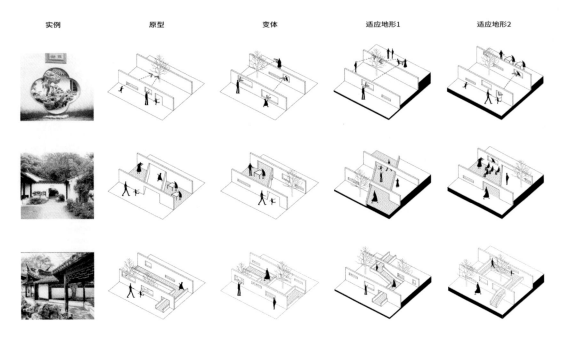

| 实例 | 原型 | 变体 | 适应地形1 | 适应地形2 |

設计题目：小径分叉的花园
指导老师：李煦
学　　生：牛冰倩

Design topic: Garden with Bifurcated Paths

Instructor: Li Xu

Student: Niu Bingqian

12

● 设计说明

方案尝试从场地周边的空间肌理和空间感受出发，以街道、围合型庭院、小广场为核心空间要素，采用建筑组群的形式，结合场地内的行进路线、视线变化和建筑的布局关系进行连接，以形成一条自北向南的完整的对外开放的街道。方案的特殊性在于将各功能空间打散，着重营造空间打散之后形成的灰空间（檐下的街道、架空的空间）以及屋顶绵延形成的屋顶平台。基于艺术之家的功能定位，这些连接空间可以承载艺术集市、艺术涂鸦以及艺术沙龙的功能，由此设计真正连接了市民生活与艺术空间，回归了城市空间功能的原真性。整体而言，街道开放于城市，艺术家独享内向型庭院，屋顶广场与市民共享，第三空间弥漫在艺术之家中，设计在建筑之间营造出了延续场地肌理的原真性巷道景观。

Design notes

The scheme attempts to start from the spatial texture and spatial feeling around the site, taking streets, enclosed courtyards and small squares as the core spatial elements, and adopting the form of building clusters,considering the routes sight change and building layout in the site to form a complete open street from north to south. The particularity of the scheme lies in the fragmentation of various functional spaces, focusing on the construction of gray space (street under eaves, overhead space) and roof continuous shape. Based on the functional positioning of art house, these linking spaces can carry the functions of art market, art graffiti and art salon. The plan truly links the citizen's life and art space, and returns to the authenticity of the urban space function. On the whole, the streets are open in the city, and the artists enjoy the introverted style. The courtyard,the roof square are shared with the citizens, and the third space is permeated in the art house. The design creates a genuine roadway landscape that continues the texture of the site between the buildings.

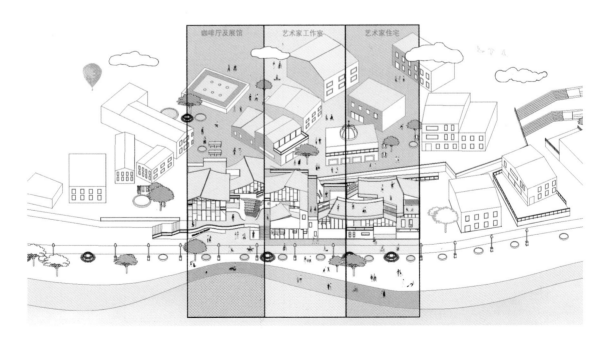

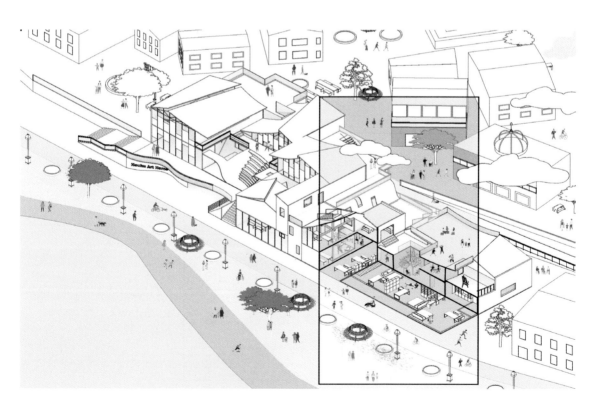

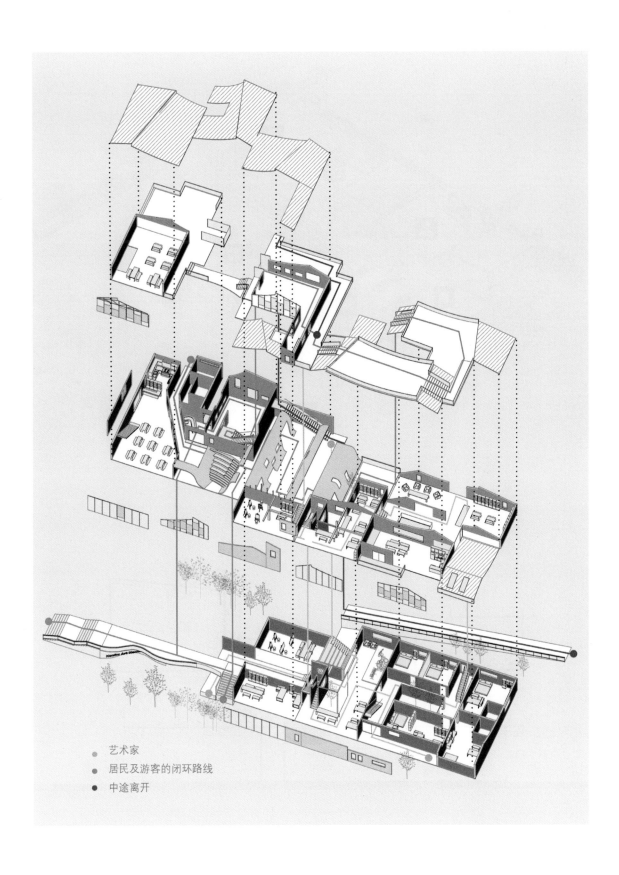

艺术家
居民及游客的闭环路线
中途离开

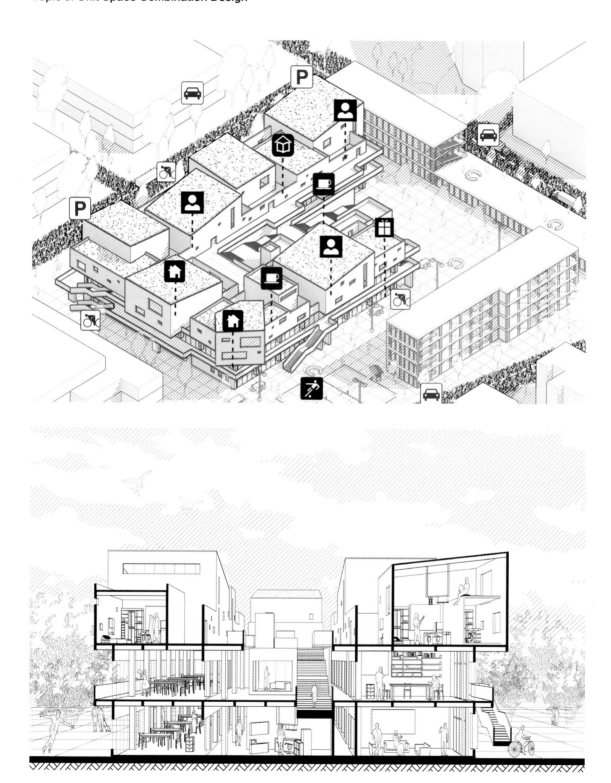

设计题目： 创客孵化器 —— 基于社区孵化项目模式的探究
指导老师： 叶强、胡赞英
学　　生： 和译

Design topic: Maker Incubator—Based on the Exploration of Community Incubation Project Model

Instructors: Ye Qiang, Hu Zanying

Student: He Yi

13

● 城市区位分析　　City location analysis

中国提出了"中国制造2025"战略，并发出"大众创业，万众创新"号召，鼓励"互联网新兴产业发展"，并制定了《"十一五"文化发展规划》《长沙市战略性新兴产业 —— 文化创意产业发展规划》等一系列政策性文件。这些科学的规划与决策的引导，为长沙市创意产业的迅速发展注入了不竭动力，帮助企业归位创新。图为长沙城市文化创意产业集聚区分布示意图。

China has put forward the "Made in China 2025" strategy and issued a call for "Mass Entrepreneurship and Innovation" to encourage the "development of emerging internet industries" according to a series of policy documents such as the "Eleventh Five-Year" Cultural Development Plan and the "Changsha Strategic Emerging Industry-Cultural and Creative Industry Development Plan". These scientific plans and decisions as guidance have injected inexhaustible impetus into the rapid development of creative industries in Changsha, helping enterprises to innovate. The following figure shows the distribution of Changsha urban cultural and creative industry cluster.

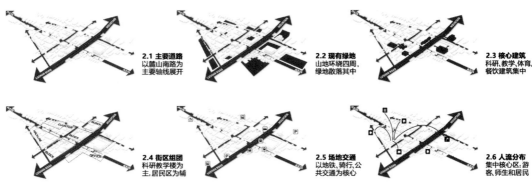

2.1 主要道路
以麓山南路为主要轴线展开

2.2 现有绿地
山地环绕四周，绿地散落其中

2.3 核心建筑
科研、教学、体育、餐饮建筑集中

2.4 街区组团
科研教学楼为主，居民区为辅

2.5 场地交通
以地铁、骑行、公共交通为核心

2.6 人流分布
集中核心区；游客、师生和居民

原有场地

现存问题：由原有入口进入场地经过两次转折。人车混流，缺少小汽车及自行车停车位

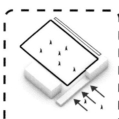

增加场地入口

解决办法：沿街立面一侧架空作为人行主入口，原有入口保留作为车行道，设停车场于一侧

退让出入口广场

绿地利用率低，轴线狭长，不符合步行尺度。解决办法：建筑后退形成入口广场，进行人群分流

休闲广场

原有绿地广场改造为活动广场。东面沿街建筑靠近居民区，附近设有居民入口

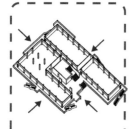

建筑作为媒介

"创客"空间是一个中间地带，连通着艺术内部对科技的延伸，也连通着日常到创意的延伸

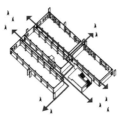

街区共享

底层开放，布置咖啡和书吧，与公共属性相契合。东南面靠近居民区，附近设有居民入口

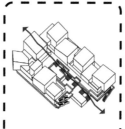

社区共享

室外大阶梯连通三层平台，共享程度逐渐降低，公私区域层层过渡，最终到达私人工作室

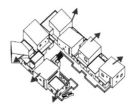

考虑场地的活动

麓山南路附近共有四类人群，调研人群日常活动时间和实际需要，在场地布置相关功能区

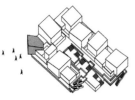

体块扭转

面向场地居民入口，与居民点进行呼应。在道路的爬升中，人们能观察场地所发生的行为

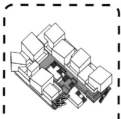

社区孵化项目模式

形成一个"组织—生产—参与—分享"机制，以便广泛地容纳参与者、分享者，让他们进入生产和传播系统中

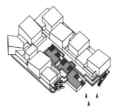

种植平台

从底层至顶层，自然景观穿插在建筑间，展示自然与人造形态的毗接

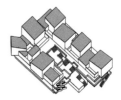

单坡屋顶

建筑坐落在岳麓山脚下，单坡屋顶与远山保持呼应，与自然对话

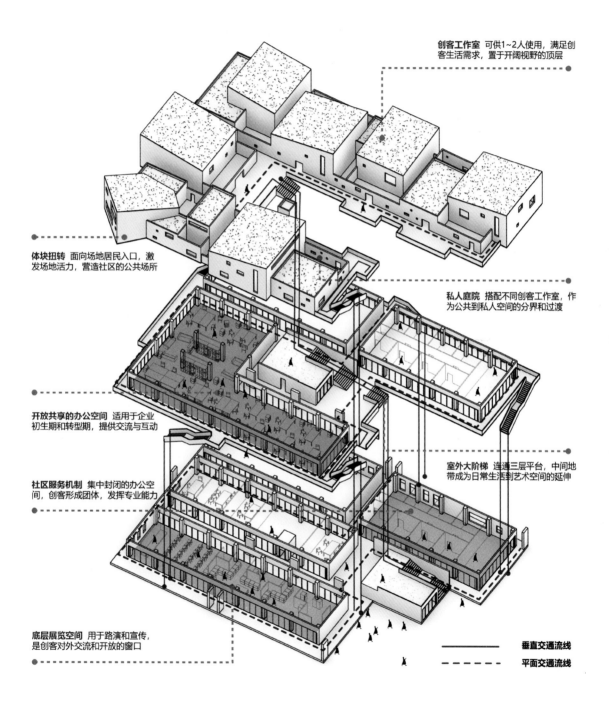

创客工作室 可供1~2人使用，满足创客生活需求，置于开阔视野的顶层

体块扭转 面向场地居民入口，激发场地活力，营造社区的公共场所

私人庭院 搭配不同创客工作室，作为公共到私人空间的分界和过渡

开放共享的办公空间 适用于企业初生期和转型期，提供交流与互动

室外大阶梯 连通三层平台，中间地带成为日常生活到艺术空间的延伸

社区服务机制 集中封闭的办公空间，创客形成团体，发挥专业能力

底层展览空间 用于路演和宣传，是创客对外交流和开放的窗口

——————— 垂直交通流线

– – – – – – – 平面交通流线

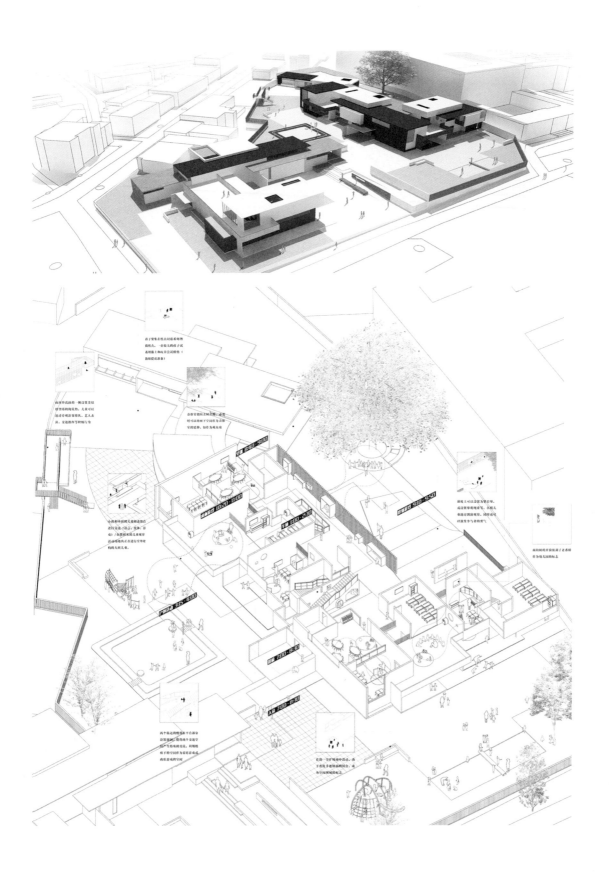

14

设计题目： THE LILIPUT—— 群体生态学下的社会模拟幼儿园
指导老师： 苗欣
学　　生： 林佳鸿

Design topic: THE LILIPUT—Social Simulation Kindergarten under Group Ecology

Instructor: Miao Xin

Student: Lin Jiahong

● **设计说明**

设想以"假装游戏"作为幼儿园的基本活动方式，将"家"作为社会最基本的一个小团体在幼儿园中体现，小朋友将以小组（可假扮家人、同事等多种社会关系）的方式共同拥有并管理一个小空间，班级以社区的形式联系多个团体。设计希望从社会结构上实现社会模拟，从而鼓励假装游戏。

● **空间要素**

以空间交叠为基本手法，将交通空间、室外领域和功能区部分交叠，使空间之间相互渗透。

用墙围合强调空间重叠，使其有同时处于两个空间的模糊感；同时功能上的交叠也能够引发丰富的交流事件。

● **空间生成**

Design notes

It is envisaged to use "pretend play" as the basic activity of kindergarten, and "home" as the most basic small group in the society to be embodied in the kindergarten. Children will be grouped (possibly pretending to be family, colleagues and other social relations).To co-own and manage a small space by connecting multiple groups in the form of community, we hope to realize social simulation from the social structure, thus to encourage pretend play.

Spatial elements

交叠部分互相连通延伸形成管道空间，管道之间通过墙面的错位和洞口发生联系形成复杂多样的流线。

通过上述关系用管道连接班级，两个班入口朝向公共灰空间，强化交流和内聚，进一步发展竖向关系。

Space generation

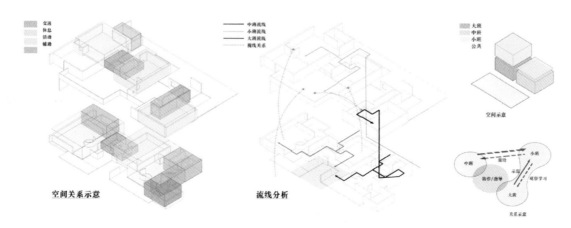

空间关系示意　　　　　　　流线分析

课题四：建筑单体强化设计
Topic 4: Strengthening Design of Single Building

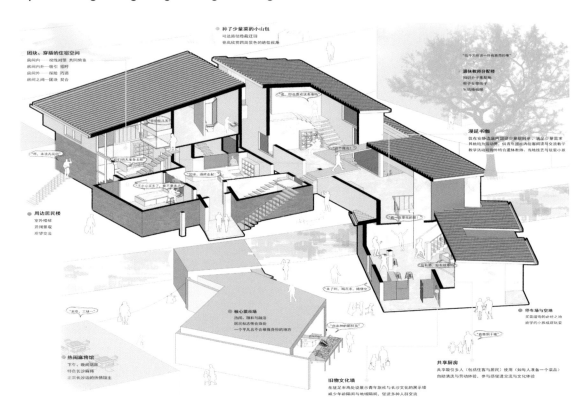

15

设计题目： 原驿 —— 国际青年旅舍设计
指导老师： 苗欣
学　　生： 陈潇、肖玉凤

Design topic: Yuanyi—International Youth Hostel Design

Instructor: Miao Xin

Students: Chen Xiao, Xiao Yufeng

● 设计说明

此次青旅课程设计针对三类人群，即旅游爱好者、背包客和青旅老板，分析其旅行触动点、意义、住宿需求、对外身份。原真性是其共同追求的核心，希冀以此达到自身的探寻与平衡。

原真有主体和客体之分，核心在于交流。我们希望通过虚面空间处理促进舒适与真诚的交流。探寻目标人群交流行为与感受，提取出空间要素进行组合。

方案选址于大学城麓山南路一类社区中心，依据场地要素，寻找青旅与社区的异同点，探索两者的融合方式。星城一隅，民风俗韵，打造的是一所怀抱社区的国际驿站。

Design notes

The design of the CYTS course is aimed at three groups of people, namely, travel enthusiasts, backpackers and hostels owners and analyzes their touching points in travel, meaning, accommodation needs, and external identity. Authenticity is the core of their common pursuit and they hopes to achieve their own quest and balance. The core of the distinction between human and object lies in communication. We hope to promote comfortable and sincere communication through virtual space processing.We explore the communication behaviors and feelings of the target group, and extract the spatial elements for combination. The plan researched in a community center of Lushan South Road in the University Town, and based on the site elements, we seek the similarities and differences between the youth hostel and the community, and explores the integration of the two. In a corner of Star City, with folk customs and charm, creates an international hotel that embraces the community.

16

设计题目: 景别叙事 —— 陈和西艺术工作室设计

指导老师: 向昊

学　　生: 谭祖斌

Design Topic: Scenery Narrative—Design of Chen Hexi Art Studio

Instructor: Xiang Hao

Student: Tan Zubin

● **设计说明**　　Design notes

陈和西

1953 年 10 月出生,祖籍湖南浏阳
湖南师范大学美术学院教授、博士生导师
中国美术家协会会员、中国油画学会理事

他的画作充满着激情,大胆的用色以及鲜
明的色块带给人活力与朝气。设计从画家
的画作出发,从三个方面将绘画空间转译
为建筑空间。

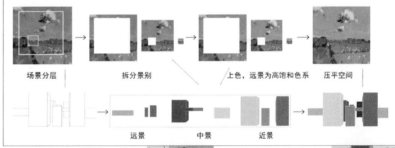

场景分层　　　拆分景别　　　上色,远景为高饱和色系　　压平空间

远景　　　中景　　　近景

景别分层

剖面　　　二层平面　　　首层平面

远景是私人的、供人远观的。对于场地而言,它是漂
浮其上的,统领场所感的,并且是艺术家居住的私人
领域。
中景是远与近的过渡区域。在公共性的首层平面,它
映射为交通;在私人性的二层平面,它代表着容纳艺
术家活动的空间。
近景是可触摸的。在剖面上,是直通三层的体块;在
首层平面上,是容纳人群特定行为的区域;在二层平
面上,是与人发生接触的家具。

色彩构成

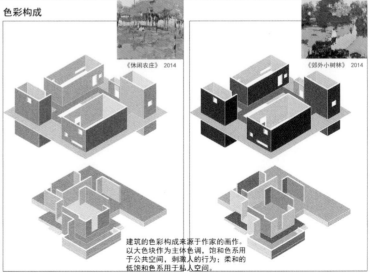

《休闲农庄》 2014

《郊外小树林》 2014

建筑的色彩构成来源于作家的画作。
以大色块作为主体色调,饱和色系用
于公共空间,刺激人的行为;柔和的
低饱和色系用于私人空间。

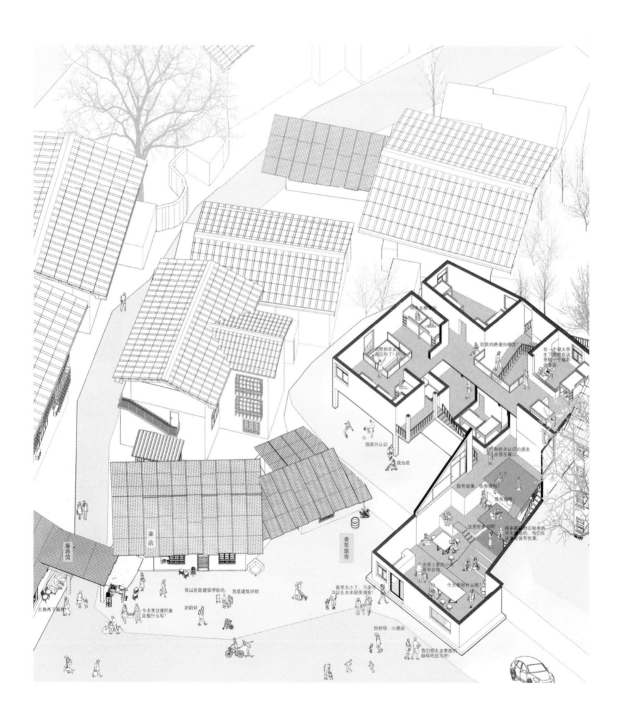

52

设计题目： "日常生活"范式的空间建构：原巷 —— 国际青年旅舍设计
指导老师： 苗欣
学　　生： 兰子千

17

Design Topic: Spatial Construction of "Daily Life" Paradigm: Original Lane—Design of International Youth Hostel

Instructor: Miao Xin

Student: Lan Ziqian

● 设计说明

当一个人爱上旅行，他在追求什么？他热衷穿行于各异的城市空间中，他追寻着陌生感和新鲜感，他渴望遇见那些有趣的灵魂，有时候走遍全世界也只为遇到那个人。

或许我们可以归结为，他在追求两种原真。一种是地域原真，他用眼睛看，用手触摸，用身体识别各个地域的气质；一种是人际原真，即开放、包容、信任等我们所追求的社会品质。

Design notes

When a person falls in love with travel, what is he pursuing? He is obsessed with those regions and is keen on traveling through different urban spaces. The sense of strangeness and freshness is what he seeks for. He longs to meet those interesting souls, and sometimes travels all over the world just to meet that person. Perhaps we can boil it down to that he is pursuing two kinds of authenticity. One is the authenticity of the region. He sees with his eyes and touches with his hands, and recognizes the temperament of each region; The other is the social qualities we pursue, such as interpersonal authenticity, openness, tolerance, trust, and so on.

● 地域原真——场地空间要素提取

由小单元组成建筑单体，建筑单体通过错层、掉层、跌落等处理适应地形。在平面上，建筑群体与路径、节点空间形成图底关系，建筑间通过夹缝、错位等方式留出路径，路径在场地内爬升、环绕或穿过建筑内部，悬空搭接，布置灵活。

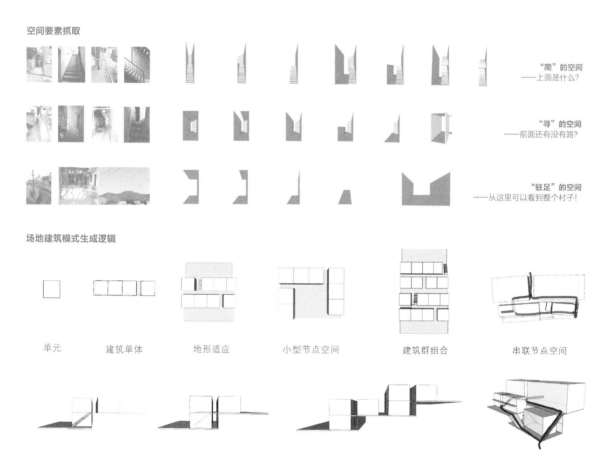

空间要素抓取

"爬"的空间
——上面是什么？

"寻"的空间
——前面还有没有路？

"驻足"的空间
——从这里可以看到整个村子！

场地建筑模式生成逻辑

单元　　　　建筑单体　　　　地形适应　　　　小型节点空间　　　　建筑群组合　　　　串联节点空间

● 地域原真——场地空间原型重构

将场地生成建筑的单元由封闭空间转化为 U 形面，利用 U 形面的导向和拒绝的双重性，在保留场地防御性特征的同时增加单元间对话、交流的可能性。

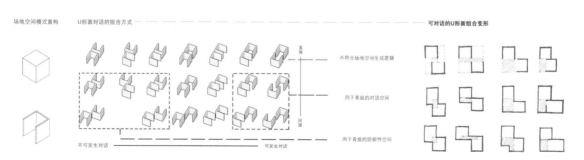

场地空间模式重构　　U形面对话的组合方式　　　　　　　　　　　　　　　　可对话的U形面组合变形

不符合场地空间生成逻辑

用于青旅的对话空间

用于青旅的防御性空间

不可发生对话　　　　　　　　　　　　　　　　可发生对话

● 人际原真——社区人群关系建构

基地位于岳麓山下一个叫"刷把冲"的城中村内，比邻
大学城和湖大附中，周边人群主要为原住村民、湖大退
休教职工、租客（大学生、低收入人群、中学生和陪读
人群）、商贩等。原住民及退休教职工内部熟络，但流
动人口与场地人际关系甚远，甚至十分有戒备心。设计
试图通过青年旅舍的置入，活化社区内部关系，使流动
人群与场地产生联系，各类人群交织，形成较为紧密的
联结。

● 人际原真——青旅人群关系建构

青旅与社区
青旅对外的"窗口"部分植入了共享厨房和咖啡厅的功能，
考虑到场地内部有大量陪读人群和青年人，共享厨房可
以成为陪读人群"逃离"封闭环境的出口，满足了其社
会交往的需求；而咖啡厅则是适合青年人社交的场所，
场地内的青年租客可以在这里自习交友。"窗口"处不
同功能的公共空间的连续性也为不同的流动人群提供了
交流的可能性。

青旅内部人群
重构后的不同开放等级的实体空间让青旅人群拥有了较
大的选择自由度，他们可以"向外"结交朋友，也可以
蜷缩在自己的私密空间而不受干扰。

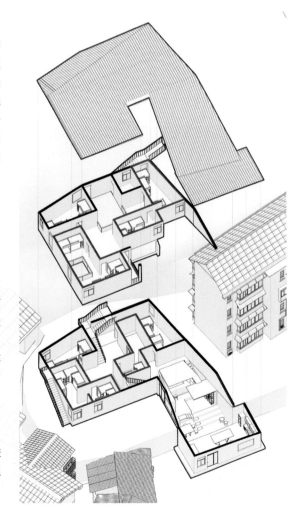

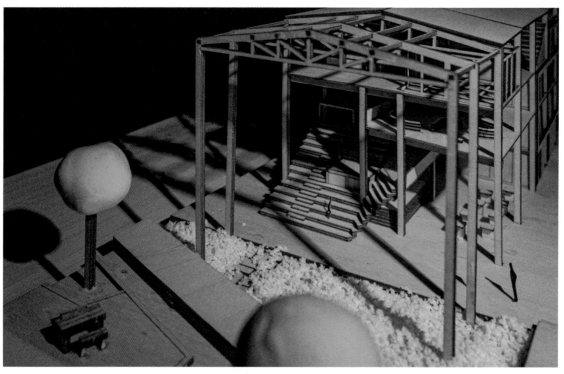

56

18

设计题目： CUBE · 媒体工作坊
指导老师： 向昊
学　　生： 冯硕

Design topic: CUBE · Media Workshop

Instructor: Xiang Hao

Student: Feng Shuo

● 设计说明

坊，一个人们耳熟能详的字眼。从古至今，它与人们的生活息息相关。设计者希望，通过无声的酝酿进行一番改造，让这间影视工作室在不破坏原有静谧环境的前提下也能够"掷地有声"，吸引更多的工作者与感兴趣者，在"校园水面"中投入一颗鹅卵石，激起文化的层层涟漪。该创意工作坊单体设计方案，源于前期设计的约 7800m²的某大学校园实习工厂空间改造规划方案。前期方案以周边的山水文化和校园文化为依托，结合场地环境，改造完成了工坊集群空间的设计。

Design notes

Workshop,a familiar word has been closely related to people's lives since ancient times. The designer hopes that, some transformations in silent generation will enable this film studio without destroying the original quiet environment to better attract more workers and amateurs to exchange ideas, just like throwing a pebble into water surface with ripples. The site of the single design plan for the creative workshop is selected from the conceptual design plan for the spatial transformation planning of an approximately 7,800 square meters in a practice factory,of university campus which was designed in the early stage.The preliminary plan relies on the surrounding landscape culture and campus culture, combined with the surrounding environment, and completes the design of the workshop cluster space.

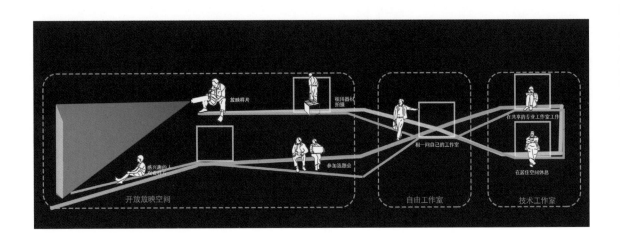

19

设计题目：红墙 · 印象 —— 艺术家联合工作室设计
指导老师：向昊
学　　生：冯永棋

Design topic: Red Wall · Impression—Artist's Joint Studio Design

Instructor: Xiang Hao

Student: Feng Yongqi

● **设计说明**

方案设计从场地人文环境及建筑与周边环境的关系出发，选择艺术家联合工作室的形式，以公共空间串联艺术家工作室单元，为艺术家与艺术家、艺术家与公众创造出公共的交流空间。红色的砖墙，作为空间序列的线索，采用地域材料，从室外的展示平台到室内的艺术画廊沿着路径流动布置，吸引并引导公众走进建筑，参与艺术活动。此外，设计还参照村落民居的形式，采用坡屋顶和室内庭院，营造丰富多样的空间，保存当地的地域记忆。现有艺术社区由部分村落民居改建而成，布局自然，砖墙为主，依稀可以见到村落的原有模样。红色的砖头是村民朴素生活的写照，也彰显了艺术家专心追求艺术的虔诚之心。

Design notes

Starting from the humanistic environment of the venue and the relationship between the building and the surrounding environment, the scheme design chooses the form of the artist's joint studio, connects the artist's studio unit with a public space, and creates a public communication space for the artist and the artist, the artist and the public. The red brick walls, as clues to the spatial sequence, use regional materials, and flow along the path from the outdoor display platform to the indoor art gallery, attracting and guiding the public into the building and participating in art activities. In addition, the design also refers to the form of village houses, using sloped roofs and indoor courtyards to create a rich and diverse space and preserve local regional memory.
The existing art community is transformed from some village houses, with a natural layout and brick walls, and you can still see the original appearance of the village. The red bricks are a portrayal of the simple life of the villagers and the devotion of the artists to pursue art.

红墙

现有艺术社区由部分村落民居改建而成，布局自然，砖墙为主，依稀可以见到村落的原有模样。红色的砖头是村民朴素生活的写照，也彰显了艺术家专心追求艺术的虔诚之心。

The existing art community is transformed from some village houses, with a natural layout and brick walls, and you can still see the original appearance of the village. The red bricks are a portrayal of the simple life of the villagers and the devotion of the artists to pursue art.

印象

艺术园区的扩展、公路的建造，让原有村落走上大拆大建道路，房子一栋栋拆毁，居民一家家离开，原有的村子记忆在其一直存在的土地上逐渐消逝。废弃的荒地被开垦种菜，原来的居民时不时回到这里，走走看看，捡拾最后的故事。

The expansion of the art park and the construction of the highway have put the original village on the road of demolition and construction, the houses have been demolished one by one, the residents have left one by one, and the memory of the original village has gradually faded away on the land where it has always existed. Abandoned wasteland is reclaimed to grow vegetables, and the original inhabitants return here from time to time to walk around and pick up the last stories.

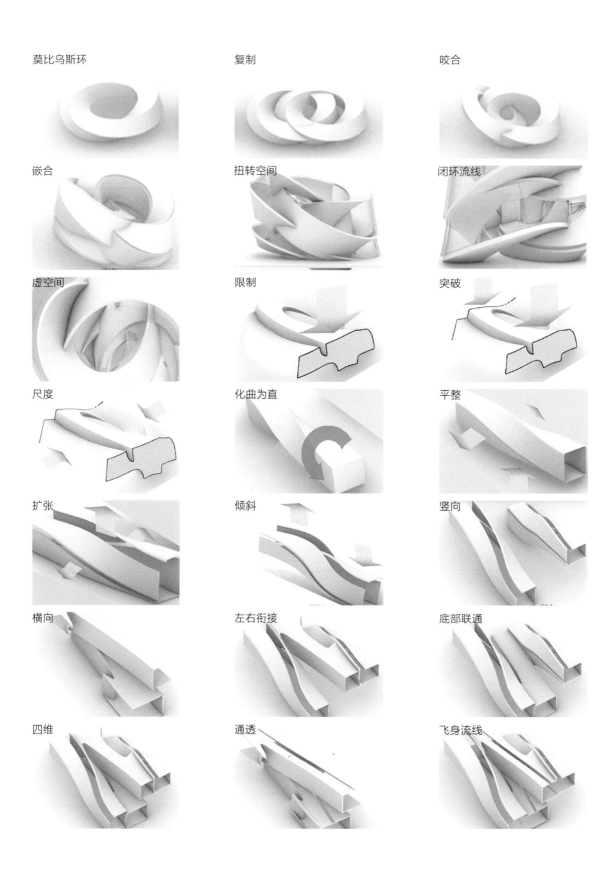

莫比乌斯环　　　　　　复制　　　　　　　　咬合

嵌合　　　　　　　　　扭转空间　　　　　　闭环流线

虚空间　　　　　　　　限制　　　　　　　　突破

尺度　　　　　　　　　化曲为直　　　　　　平整

扩张　　　　　　　　　倾斜　　　　　　　　竖向

横向　　　　　　　　　左右衔接　　　　　　底部联通

四维　　　　　　　　　通透　　　　　　　　飞身流线

设计题目： FIRE LOOP| 赛车研发俱乐部设计
指导老师： 谢菲
学　　生： 丁一凡

20

Design topic: FIRE LOOP | Design of Racing R&D Club

Instructor: Xie Fei

Student: Ding Yifan

● 设计说明

工业对城市生态和现代文化的交互关系一直是城市设计者们的永恒话题。设计方案选址在某大学城校园内的工业遗产保护和创新中心改造地块上。设计敏感地把握到校园内现代科技机构及产业所具有的先锋空间语汇，通过富有戏剧性的方案表达，来展现未来城市社会应该具有的精神面貌。

设计方案还结合新兴的智能建造、3D 打印建筑空间等要素，思考可持续技术的应用；应用参数化等智能语言编译空间几何，来创建高速、节能、智慧、充满活力与未来性的智能赛车创新群体所偏好的研究交流环境。设计方案个性鲜明，场所营造充满热情，空间氛围有视觉冲击力，体现了设计者视角下对无限美好未来的富有感染力的独特解读。

Design notes

The interaction between industry and urban ecology and modern culture has always been an eternal topic for urban designers. The design plan is located on the industrial heritage protection and innovation center transformation plot of a university city campus, sensitively grasping the pioneering spatial vocabulary of modern scientific and technological institutions and industries on the campus. The spiritual appearance that future city should have is expressed through dramatic plan expressions. The design plan also considers the application of sustainable technology in combination with elements such as emerging intelligent construction and 3D printing building space, and uses intelligent languages such as parameterization to compile spatial geometry to create a high-speed, energy-saving, intelligent, energetic and futuristic research communication environment which intelligent racing innovation group preferred. The design plan has distinctive characters and the place is full of enthusiasm, and the space atmosphere has a visual impact, which reflects the unique and infectious interpretation of the infinitely beautiful future from the designer's perspective.

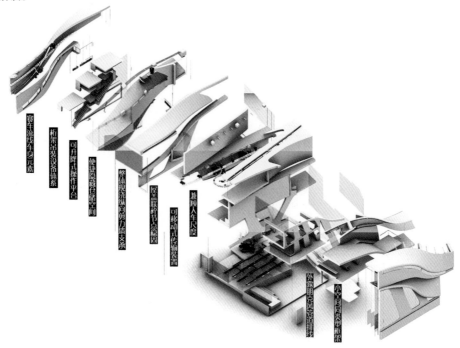

设计题目： "破茧"——新信息乌托邦的营造 | 校园工坊单体设计
指导老师： 谢菲
学　　生： 覃艺

21

Design topic: "Broken the Cocoon"—the Creation of a New Information Utopia | Campus Workshop Single Design

Instructor: Xie Fei

Student: Qin Yi

● **设计说明**

本设计从营造新的信息乌托邦这一愿景出发，考虑原基地的超算技术优势和跨学科人才优势，并基于第一轮游牧式创意社区组团设计，设置一个由数据挖掘、传播分析、媒体发声三大团队组成的舆情应对创研工坊。

方案适度打破了已有舆情研究团队的工作模式，由此细化了空间需求。同时，由茧洞的空间形态确定了独立式窑洞的原型及筒壳结构初选型，并且尝试将原型打破，将其组织成虚实、自然、光影和谐的理想空间。希望在这隐喻着破茧新生的空间里，创研团队能对信息时代及其茧洞效应有更多的思考。

Design notes

This design starts from the desire to create a new information utopia, considering the advantages of supercomputing technology and interdisciplinary talents of the original base.And based on the first round of nomadic creative community group design, we set up a workshop of creation and research for public opinion composed of data mining,dissemination analysis and media voice the three major teams.The plan moderately breaks the existing working mode of research team, , thereby refines the space requirements. At the same time, the prototype of the free-standing cave dwelling is determined by the spatial form of the cocoon cave, as well as the preliminary selection of the cylindrical shell structure, and tries to break the prototype to organize it into an ideal space of virtuality and reality, nature, light and shadow in harmony. It is hoped that in this metaphorical space for new life, the innovation team can think more about the information age and its effects.

Q1: 打破常规的舆情创研工坊可以有哪些场地潜力？

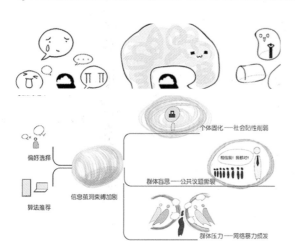

在信息生活愈发自由的时代，我们几乎能选择"我想知道的一切"。但，日益强大的推荐算法会不会把海量的言论过滤得只剩"我爱听的声音"？长此以往，我们是否会被缚于茧洞之中？

若信息茧洞不可避免，那我们该如何打破桎梏?尤其在舆情风向方面，能否成立新的舆情创研工坊来破解茧洞弊端?能否营造新的信息乌托邦以适应创研团队需求？

63

设计题目： 气味重启计划
指导老师： 余燚
学　　生： 蔡永怡

22

Design topic: Scent Restart Plan

Instructor: Yu Yi

Student: Cai Yongyi

● **设计说明**

气味能够唤起人的记忆，建立起强烈的、关联性的、丰富的时空间，赋予建筑"电影式"的功能。该建筑定位为中式烘焙师工作室，中式糕点特有的强回忆性气味能够造成时空无限延伸，使人产生对以往记忆的更多思考和探寻。本设计欲利用这一特点，对气味进行引导，从而使空间使用者在糕点香味中回溯记忆，使该建筑成为维系周边居民与师生、集体与个人、过去与现在的空间纽带。

Design notes

Smell can evoke people's memory, establish a strong, relevant and rich space and time, and give architecture a "cinematic" function. The building is positioned as a Chinese baker's studio. The unique strong reminiscent smell of Chinese pastries can cause infinite extension of time and space and make people think and explore more about past memories. This design wants to use this feature to guide the smell, so that the space users can recall the memory in the cake flavor, and make the building become a spatial link between the surrounding residents and teachers and students, the collective and the individual, past and present.

Shape Generation—形态生成

空间体块	功能分割	气流处理	水平错层	立面处理
根据红线范围限定34m×12m×12m的空间体块	整理烘培师工作生活和顾客用餐的空间需求，将狭长体块分为A、B两部分	A栋置入风筒，通过热动力气流运动，促使扩香；B栋由圆形中庭自然散香	通过水平错层使工作室拥有漫反射光线的天窗，起居室形成朝南阳台	引入百叶窗系统构成空间边界限定，用以调节气流和室温，丰富光影体验

Traffic Analysis—交通流线分析图

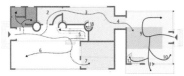

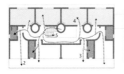

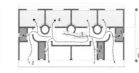

1: 入口-厕所 6: 休闲客厅 7: 餐厅-厨房
5/12: 上楼 8: 旋转楼梯 9/10: 选购-付款-用餐
2-3-4-9: 消毒更衣-烘培坊-送餐通道-糕点展销

1/9: 楼梯-厕所 2/3: 起居室-阳台
4: 楼梯-工作室 5: 上楼 6: 交通核
7: 楼梯-DIY体验室 8: 楼梯-茶室

1: 楼梯-厕所 2/3: 起居室-阳台
4: 楼梯-工作室 5: 到达景观平台

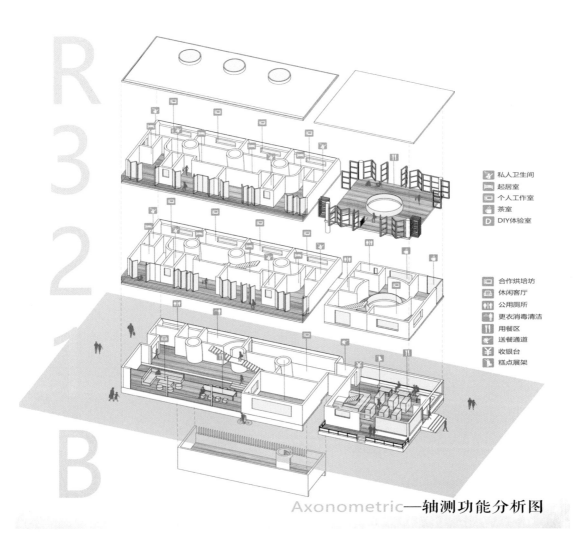

私人卫生间
起居室
个人工作室
茶室
DIY体验室

合作烘培坊
休闲客厅
公用厕所
更衣消毒清洁
用餐区
送餐通道
收银台
糕点展架

Axonometric—轴测功能分析图

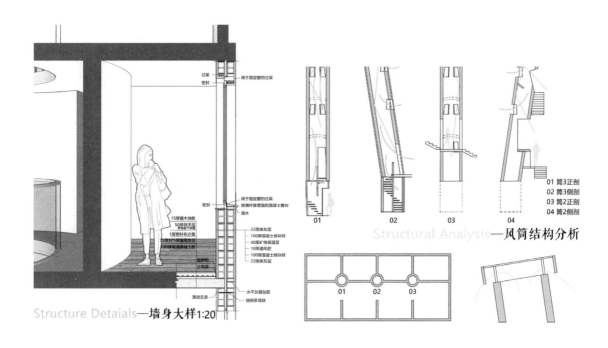

过梁
密封 用于固定窗的过梁

用于固定窗的过梁
密封
玻璃纤维增强的混凝土窗台
滴水

15厚铺木地板
50厚找平层
1厚塑料布分离
15厚EPS保温隔音
+30厚垫渣混凝土层
踢脚板
分离垫

22厚抹灰层
100厚混凝土砌块砖
40厚矿物保温层
10厚通风腔
100厚混凝土砌块砖
22厚抹灰层

水平灰缝加筋
滑动支承
铬钢系墙铁

Structure Detaials—墙身大样1:20

Structural Analysis—风筒结构分析

01 筒3正剖
02 筒3侧剖
03 筒2正剖
04 筒2侧剖

01   02   03   04

01   02   03

Airflow analysis—气流分析

LIVING   WORK
LIVING   WORK
KITCHEN   WORK
STORE

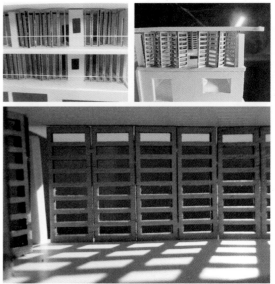

三年级
Third grade

三年级建筑设计课程介绍
Course Introduction of Architectural Design Course for Grade Three

课程内容：建筑设计 III 、建筑设计IV

Course content: Architectural Design III, architectural Design IV

教师团队
Teacher team

蒋甦琦	陈翚	李旭	彭智谋	张蔚
Jiang Suqi	Chen Hui	Li Xu	Peng Zhimou	Zhang Wei

龚震西	罗荩	许昊皓	张光	吕瑞杰
Gong Zhenxi	Luo Jin	Xu Haohao	Zhang Guang	LYu Ruijie

课程介绍
Course introduction

本年级的课程是建筑学专业本科生的主干专业课程，包括建筑设计 III 和建筑设计 IV。近 5 年来，三年级课程教学组把设计课程的教案编制作为课程建设的重要内容，系统而持续地编制了三年级设计类课程教案，取得了较好的阶段性成果。教学组织上，规定性选题由全年级组织，参加教学指导的老师一般超过 8 位。在教学实践中发现，由于教师执教能力、学科方向的差异和年轻教师的培养等原因，集体性教案的编制尤为重要。同时，在全年级推进的教学中，也鼓励部分教师作独立性小组教学探索，如数字化方法为导向的课程设计；同样，对于专题类的课程设计而言，确保既有专题导向，又积极鼓励国际化、多样化、研究型选题。

在教学执行计划中，大课的组织尤为关键。整个学期安排 8 次共 16 学时的讲课，按各个阶段的教学内容和知识点，组织 8 位教师主讲各个相关主题，大课的主讲老师不局限于该课程的指导教师，从课程的需要出发，邀请建筑系各学科团队中有研究专长的教师参与教学。同时，在技术深化环节，从设计院和专业公司聘请多位资深技术人员担任课程顾问，充分发挥大课教学方式在知识传授的系统性、完整性和即时性方面的优势。大课的内容

This year's courses are the main professional courses for undergraduates majoring in architecture, including architectural design III and architectural design IV. In the past five years, the teaching group for the third grade has taken the preparation of teaching plan of curriculum design as an important part of curriculum construction, and has continuously prepared the systematical teaching plan of the third grade curriculum design, achieving good phased results. In terms of teaching organization, the specified topics are organized by the whole grade, and generally more than 8 teachers participate in teaching guidance. In teaching practice, it is found that the preparation of collective teaching plan is particularly important due to the differences in teachers' teaching ability, discipline direction and the training of young teachers. At the same time, in the teaching process covering the whole grade, a small number of teachers are encouraged to explore independent group teaching, such as digital method-oriented curriculum design. Similarly, for the curriculum design of special topics, we need to ensure the existing project orientation and actively encourage the exploration of international, diversified and research-based topics.

In the teaching implementation plan, the organization of large courses is particularly critical. There are 16 lessons which are divided into 8 times in the whole semester. According to the teaching contents and knowledge points of each stage, eight teachers are organized to give lectures on various related topics. The main lecturers of large courses are not limited to the instructors of the course. On the contrary, teachers with research expertise in the discipline team of the Department of Architecture are invited to

包括基本原理，如公共建筑设计原理、历史保护类建筑设计原理、基于结构创新的建筑设计等；设计方法，如集合性商业设施的空间组合、生态技术的数字设计方法、复杂城市环境中的行为学等；设计深化与表达，如设计制图的表现与实现、基于虚拟现实的建筑体验表达设计等，有效提升学生的设计深化能力。

建筑设计 III 课程简介

建筑设计 III 是建筑学专业本科生的主干专业课程。该课程通过讲授在城市背景下中小型综合性公共建筑设计基本理论及知识，要求学生熟练掌握城市尺度与人文环境下的中小型综合性公共建筑的设计方法。主要培养学生掌握常用的平面功能组织和空间设计的方法。以"场所场域""复合行为""空间营造""材料与建构"等为出发点，要求熟悉人群行为的观察与记录，把握建筑设计概念生成过程及建筑设计思维方法，掌握基本空间逻辑的推演与生成过程，了解基本的材料与建构的方法，并掌握建筑设计的表达与表现方法。

教学目标

学习和掌握中小型公共建筑基本设计方法，理解城市背景下中小型综合性公共建筑的空间、材料、形式等与场所的关系。学习从场地调研与人群观察中发现设计问题，并最终找到解决问题的方案。要求熟悉人群行为的观察与记录，把握建筑设计概念生成过程及建筑设计思维方法，掌握基本空间逻辑的推演与生成过程，了解基本的材料与建构的方法，并掌握建筑设计的表达与表现方法。在设计过程中通过实地调研、工作模型等方式，重点加强学生制作工作模型的能力，对中小型综合性公共建筑类型的理解以及空间组合方式原理的学习。通过对基地的实地调研，加强对学生设计的整体环境观的培养。同时通过快速设计的训练，加强学生快速构思以及草图表达的能力。本课程的教学思政目标为培养学生在复杂城市背景下理解建筑的环境与人文关怀问题。

建筑设计 IV 课程简介

建筑设计 IV 是建筑学专业本科生的主干专业课程。该课程通过讲授中型文化类和居住综合类建筑设计的基本理论及知识，要求学生熟练掌握城市中复杂条件下的中型文化及居住综合类建筑的设计方法。本课程设计旨在锻炼设计者对场地环境敏锐的认知能力，将环境要素与建

participate in the teaching process based on the needs of the course. At the same time, in the deepening process of technology, a number of senior technicians are employed as course consultants from design institutes and professional companies to give full play to the advantages of large-course teaching method in the systematicness, integrity and timeliness of knowledge transfer. The contents of large courses include basic principles, such as the design principles of public buildings and historical protection buildings, and architectural design based on structural innovation, etc.; design methods, such as spatial combination of collective commercial facilities, digital design method of ecological technology, and ethology in complex urban environment; the deepening and expression of design, such as the expression and realization of design drawing, architectural experience expression design based on virtual reality, etc. All of these can effectively deepen students' design ability.

Introduction to Architectural Design III

Architectural Design III is the main professional course for undergraduates majoring in architecture. By teaching the basic theory and knowledge of the design of small and medium-sized comprehensive public buildings in the urban context, the course requires students to master the design methods of small and medium-sized comprehensive public buildings in the urban scale and humanistic environment. It mainly trains students to master the commonly used methods of plane functional organization and space design. Starting from "place and field" "composite behavior" "space construction" and "materials and construction", students are required to be familiar with the observation and record of crowd behavior, grasp the generation process of architectural design concept and architectural design thinking method, master the deduction and generation process of basic spatial logic, and understand the basic materials and construction methods, and acquire the methods of expressing architectural design.

Teaching objectives

Learn and master the basic design methods of small and medium-sized public buildings, and understand the relationship between building space, materials, forms and places of small and medium-sized comprehensive public buildings in the urban context. Learn to find the problems of design through site investigation and crowd observation, and finally find solutions to the problems. Students are required to be familiar with the observation and recording of crowd behavior, grasp the generation process of architectural design concept and architectural design thinking method, master the deduction and generation process of basic spatial logic, understand the basic materials and construction methods, and acquire the methods of expressing architectural design. In the design process, through field research, working model and other methods, we focus on strengthening students' ability to make working model, and their understanding of small and medium-sized comprehensive public buildings and to learn the principle of space combination. Through the field investigation of the base, strengthen the cultivation of students' overall environmental view of design. At the same time, through the training of rapid design, we will strengthen students' ability of rapid conception and expressing by sketching. The teaching ideological and political goal of this course is to cultivate students' understanding of architectural environment and humanistic issues under the background of complex cities.

筑空间巧妙融合进行设计。在解决内部空间和功能等问题的同时，充分体现文化及居住综合类建筑的场所精神。要求学生对场地基本物质环境条件、场地的历史文脉、城市空间条件和使用状况以及中型文化及居住综合类公共建筑的功能、空间、形式等方面有熟练掌握。深化建筑设计概念生成过程及建筑设计思维方法，熟练掌握基本空间逻辑的推演与生成过程，深入认识文化及居住综合类建筑的结构形式，以及光、材料及建构方式对建筑和空间、细部产生的影响，并了解人文与生态环境方面的知识。

教学目标

锻炼设计者对场地环境敏锐的认知能力。掌握调研、收集资料的科学方法，了解历史建筑改造和再利用的设计方法，提高处理较复杂环境问题和设计构思及方案表达的能力，最终使学生具备中型建筑方案设计的能力。要求对场地基本物质环境条件、场地的历史文脉、城市空间条件和使用状况以及中型文化及居住综合类公共建筑的功能、空间、形式等方面有熟练掌握。深化建筑设计概念生成过程及建筑设计思维方法，熟练掌握基本空间逻辑的推演与生成过程，深入认识文化及居住综合类建筑的结构形式，以及光、材料及建构方式对建筑和空间、细部产生的影响，并了解人文与生态环境和建筑节能方面的知识。本课程的教学思政目标为培养学生在历史文脉背景下理解建筑的遗产属性以及文化传播。

Introduction to architectural design IV
Architectural design IV is the main professional course for undergraduates majoring in architecture. By teaching the basic theory and knowledge of medium-sized cultural and residential comprehensive architectural design, the course requires students to master the design methods of those buildings under complex conditions in cities. This course is designed to train the designer's keen cognitive ability of the site's environment and skillfully integrate the environmental elements with the architectural space. While solving the problems of internal space and function, it fully embodies the place spirit of cultural and residential comprehensive buildings. Students are required to master the basic material and environmental conditions of the site, the historical context of the site, urban space conditions and use conditions, as well as the functions, spaces and forms of medium-sized cultural and residential comprehensive public buildings. Deepen the generation process of architectural design concept and architectural design thinking method, master the deduction and generation process of basic space logic, deeply understand the structural form of cultural and residential comprehensive buildings, as well as the impact of light, materials and construction methods on the details of architecture and space, and understand the knowledge of humanities and ecological environment.

Teaching objectives
Exercise the designer's keen cognitive ability to the site's environment. Master the scientific methods of investigating and collecting data,
understand the design methods of transforming and reusing historical buildings, improve the ability to deal with more complex environmental problems, and improve their ability to propose design ideas and schemes, and finally enable students to have the ability to design medium-sized buildings. Students are required to master the basic material and environmental conditions of the site, the historical context of the site, urban space conditions and use conditions, as well as the function, space and form of medium-sized cultural and residential comprehensive public buildings. Deepen the generation process of architectural design concepts and architectural design thinking methods, master the deduction and generation process of basic space logic, deeply understand the structural forms of cultural and residential comprehensive buildings, as well as the impact of light, materials and construction methods on buildings, spaces and details, and understand the knowledge of humanities and ecological environment and building energy conservation. The teaching ideological and political goal of this course is to train students to understand the heritage attribute of architecture and cultural communication under the background of historical context.

后浪时代的大学空间

组织人：彭智谋

2020 年湖南省大学生可持续建筑设计竞赛

Topic 1: University Space in the Post Wave Era

一、竞赛主题

本次竞赛以"后浪时代的大学空间"为题，旨在考查同学们以校园空间的建设者与参与者的身份，如何批判性思考大学校园空间需求的本质、技术与人的关系；又如何以设计思维和技术手段去解决现实社会与生态问题，使之成为延续和传播校园物质文化的重要载体，达到以文"化"人、环境育人的教育效果。

二、设计内容及要求

1. 设计内容

本次设计自选大学校园内的真实场地，范围不限。可选择的类型如下：

（1）校园新建筑设计：在已有校园空间自选场地，设计3000~6000m²的校园建筑，内容和功能自定。

（2）校园老建筑改扩建设计：在某个已建成的老校区内选择一个3000~6000m²的建筑空间进行改造设计，内容和功能自定。

2. 设计要求

（1）参赛方案的真实场地要求：所有方案应采用实际地形，立足于调查研究，在理性分析的基础之上进行设计，

体现出研究型设计的特点。

（2）提交方案包括但不限于规划、建筑空间设计、交互设计等。我们鼓励跨学科、跨专业的设计合作，以期促成不同领域的深度合作，创新设计与数字技术在未来校园的深度应用及合作，真正实现对于未来校园的创新性改变。

（3）图纸表达规范：图纸能充分表达作品创作意图，且需包含必要的设计说明（可组合于图面之中）等，比例不限；竞赛官方语言为中文，度量单位为公制单位。

（4）要求采用通用建筑设计软件绘制成图。鼓励但不强制使用点云数据导入 Autodesk Revit 软件处理生成BIM 模型。

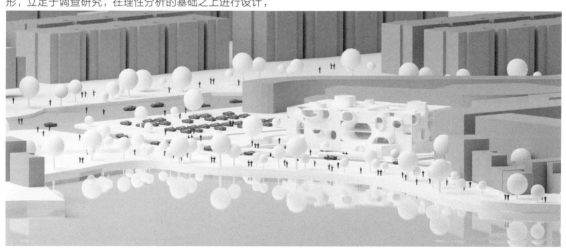

《情绪博物馆——我与黑猫》和译

课题二

长沙凯雪面粉厂遗址改造与更新
Topic 2: Reconstruction and Renewal of Kaixue Flour Mill Site in Changsha

组织人：罗荩

一、设计主题

长沙凯雪面粉厂位于长沙市城北开福区潘家坪路、幸福桥附近，处于黄兴北路、开福寺路、潘家坪路的交叉地带。基地内有大量20世纪遗留的面粉厂厂房遗址建筑。厂区位于城市居民区，道路景观均有过规划设计，场地基础条件较好，建筑具有保护、改造再利用的价值。

本课程设计旨在锻炼设计者对场地环境敏锐的认知能力，并将环境要素与建筑空间巧妙融合进行设计。在解决内部空间和功能等问题的同时，充分体现博览建筑的"场所精神"。

二、设计内容

1. 规划总平面图

在原有厂区范围内，完成城市文化创意园规划总图的策划及总平面图设计。

文化创意艺术园总建筑面积为25000m²（±5%），应至少包含各功能分区（可根据拟设计的主题博物馆进行其他功能区特色布置，需在原有厂房总图基础上，规划出各个功能区所需面积）

2. 单体博物馆设计

（1）改建建筑面积：1000~3000m²

（2）新建建筑面积：2500~3500m²

（3）4、5、6、7、8、9栋至少保留3栋（可考虑6、9栋筒仓至少保留1栋）。

（4）需考虑与文创园规划总平面图的衔接，适当结合室内外景观设计。

三、设计要求

1. 要求梳理整个项目的策划定位、图底关系，营造丰富的室内外空间。除拆除和新建部分外，其余均在原有总图基础上布局，并对厂区适当更新改造，激活原有场所，创造新的景观环境。

2. 各组在基地范围内选取合适的地块，进行改、扩建设计。所选择地块内厂房等原有建构筑物可拆除、保留、新建。将保护和适应性改造有机结合起来，原厂房作为新建筑的一部分功能空间载体，对其进行改造，并和新建筑一起进行综合设计。

四、地块划分

基地红线范围内，数字楼房为可保留或者改造的建筑（4、5、6、7、8、9栋至少保留3栋，可考虑6、9栋筒仓至少保留1栋），需结合规划道路考虑出各入口方向。

凯雪面粉厂现场照片

城市慢生活：社区图文信息中心设计

组织人：罗荩、李旭

Topic 3: Urban Slow Life: Design of Community Graphic Information Center

一、设计主题

本课程设计从城市—社区—建筑的层面，尤其是从社区环境这一中观层面，探讨建筑与城市空间环境的关系；从建筑设计的角度，研究建筑设计空间对于社区环境、城市有机空间的意义。学习和掌握中小型公共建筑的基本设计方法，理解建筑设计形式、空间，材料及其与环境的关系。

基地选址位于黄兴北路和湘春路交界处片区，选择A、B两个地块展开设计。地块A位于黄兴北路老城街区，湘春路以南片区，民主巷片区。地块B以长沙市基督教城北堂为中心，位于黄兴路以东、湘春路北侧地块。地块以长沙市基督教城北堂为中心。

二、设计内容

图书馆总建筑面积为420010m²。由阅览、书库、借阅、目录检索、学术报告厅、内部办公及技术用房、培训教室、公共活动用房等部分组成。具体内容及规模设置如下：

阅览室（藏书、阅览合一）：1200m²

基本书库：200m²（书库应靠近借阅处）

借阅与目录检索：140m²

学术报告厅200座：300m²（包括设备用房15m²、休息接待室20m²）

研究室100m²

设置独立的专用阅览室6~8间（15m²）

内部技术用房及办公：140m²

教学培训用房：330m²

社区公共活动用房：为社区居民及市民提供一定的公共活动空间，可结合交通空间设置。

综合性活动用房：共600m²

辅助面积：1200m²（其中包括门厅、厕所、走道、楼电梯间等）

三、设计要求

1. 设计重点落在人的行为与感知、日常生活体验；空间叙事、空间逻辑与概念生成；空间结构、空间边界、空间路径；材料、结构与建构细部；光影与室内外空间、景观与环境。

2. 设计难点

（1）通过现场记录、影像剪辑，进行空间叙事，完成抽象空间绘本。

（2）延续城市肌理，打破均质的城市肌理。

（3）活化城市和社区的边界空间。

（4）通过环境行为认知，抽取设计概念，营造场所与空间。

3. 设计成果

（1）城市空间绘本以及设计手工模型：概念草模

（2）设计图纸：A1图纸3~6张

总工会现场照片

课题四

未来：我的大学
2020 年第十四届谷雨杯设计竞赛
Topic 4: Future: My University

组织人：蒋甦琦

一、设计主题

"未来：我的大学"该主题立足于当代中国大学校园生活的现实并指向未来。对于"我的大学"之未来畅想，实质上是回应当今校园中的问题，集结技术和思想的力量，去创造一个更加多元、智慧、可持续的大学校园环境。

关键词：数字技术、可持续、＿＿＿（根据自己的设计置入）

二、设计内容

本设计可以是新校园建设或校园老建筑改建的一部分，需要容纳教学、生活、生态空间三个部分。可选择的类型：

（1）校园新建筑设计：在已有校园空间自选场地，做一个 4000~6000m² 的建筑设计，内容和功能自定。

（2）校园老建筑改建设计：选择一个 3000~5000m² 的校园老建筑空间进行改造设计，内容和功能自定。

三、设计要求

（1）参赛方案的真实场地要求：所有方案应采用实际地形，立足于调查研究，在理性分析的基础之上进行设计，体现出研究型设计的特点。设计图纸应包括现场照片及实地调研与分析的内容。

（2）建筑形态及构建形式的数字化：能够充分体现 BIM 软件的功能与作用，参赛作品鼓励使用采集的点云数据导入 Autodesk Revit 软件处理生成 BIM 模型，鼓励运用 VR 虚拟现实生成软件在 BIM 环境（Revit 平台）中一键发布可漫游、可交互的虚拟场景执行程序。如经技术审查确认参赛作品没有提供 BIM 模型，将直接取消其参赛资格。

（3）提交方案：包括但不限于规划、建筑空间设计、产品设计、交互设计等。我们鼓励跨学科、跨专业、跨国家或地区的设计合作，以期促成不同领域的深度合作，创新设计与数字技术在未来校园的深度应用及合作，真正实现对于未来校园的创新性改变。

《浮于事》宋智明、王昊、方万俊、庞艳

捷克 Liberec 博物馆改扩建设计

组织人：陈犟

Topic 5: Reconstruction and Expansion Design of Liberec Museum

一、设计主题

本次课程以"博物馆扩建"为主题，旨在考查学生把握新老建筑之间关系问题的能力，鼓励学生以空间建设者的身份，探索展览空间需求的本质，系统地理解城市建筑，以设计思维和技术手段去解决现实社会问题，使之成为延续和传播物质文化的重要载体；掌握博览建筑设计的一般原理与方法，促进学生对地方文化的理解，体会建筑与所处环境之间的密切关系，探索提炼、运用地域建筑风格进行创作的路径，表达一种独特的文化内涵，进一步培养学生在建筑空间与意境表现方面的创造力。

二、设计内容

1. 概况简介

利贝雷茨（Liberec）市位于捷克尼萨河畔，被伊泽拉山脉包围，是捷克第四大城市，为利贝雷茨州的首府和最大城市。这座城市的历史可以追溯至 14 世纪，经过多个世纪的发展，已经成为一座风景如画的古城。

波希米亚博物馆位于利贝雷茨老城区的郊区，在 Vitezna 和 Masarykova 街的交叉处 Masarykova 11 号（见区位图）。这座博物馆由 F. Ohmann 设计，建于 1897~1898 年，距今约 124 年。到目前为止，该建筑没有进行过任何重大的改造，建筑整体被完整保存下来。然而，这也导致了今天的博物馆在建筑功能方面遇到了问题，特别是在保持展览空间和仓库的稳定气候条件方面，同时游客公共空间也相对缺乏。

2. 面积要求

本课题包含波希米亚博物馆的内部改造与扩建设计，要求总面积不低于 5000m²，扩建范围位于博物馆东北区域，主要空间组成及其参考使用面积要求如下（可根据设计构思适当增设，要求考虑无障碍设计）：

（1）新的展览空间，最少 2300m² 的展览空间，分为 4 个部分。

（2）仓库：最少 300m² 的存储空间。储藏室必须与一个面积至少为 25m² 的研究室相连。研究室仅供博物馆馆长及研究人员使用，房间必须有自然光。

（3）技术修复车间：面积不小于 250m²。

（4）游客服务空间面积自设。

（5）员工空间面积自设。

三、设计要求

1. 充分考虑当地气候条件和地域特征，结合对当地建筑的认知、理解与探索，表达出一定设计思维。

2. 功能分区合理，展览区、仓库区、技术和办公用房及游客服务设施等部分有机结合。

3. 空间流线设计合理，具连续性与灵活性，注意内外空间组织，合理安排休息空间。

4. 把握新老建筑之间的关系，充分考虑新老建筑之间的过渡连接空间设计。

5. 考虑周边地貌及建、构筑物对其建筑群体的视觉关系，可选择地方传统材料和做法与现代结合，创造丰富的外观效果。

6. 鼓励积极探索适应时代发展要求的、富于创造力的设计方案。

7. 图纸表达规范：图纸能充分表达作品创作意图，且需包含必要的设计说明（可组合于图面之中）等，比例不限。

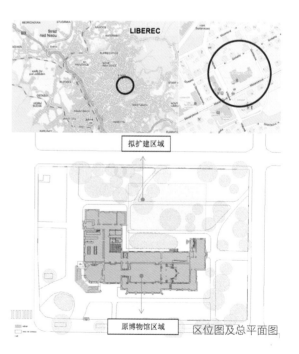

区位图及总平面图

课题六

汉口胜利仓库改扩建设计

组织人：张蔚

Topic 6: Reconstruction and Expansion Design of Shengli Warehouse in Hankou

一、设计主题

本次课程以 2018 届四校联合毕业设计题目为背景，对汉口原德租界德瑞洋行仓库/胜利仓库进行改扩建设计，旨在将保护和适应性改造有机结合起来，原物资仓库作为新建筑的一部分功能空间载体，对其进行改扩建；并对其进行结构分析、空间改造、功能置换、立面改造等。结合周边街区局部的城市更新空间改造，将之与改扩建部分一起进行综合设计。注重新老城市空间和建筑的协调与对话、空间的衔接过渡等。

二、设计内容

设计以体现城市记忆、工业文明景观、市民生活等为主题的博物馆、美术馆等展示类文化中心建筑、文化创意中心或其他文娱休闲类建筑，设计者经过调研确定展示内容和设计命题，例如"城市记忆博物馆""现代艺术博物馆""近现代租界文化博物馆"等。将此处建成一个集展览、教育、休闲为一体的城市博物馆文化场所空间。

三、设计要求

1. 建筑设计要求

（1）场所营造

①通过建筑设计和外部环境设计，营造符合基地特有文化主题的场所，以体现博览建筑较高的文化和艺术内涵。

②尊重环境：建筑应与环境密切结合，使建筑群体融入环境中，使建筑内外环境产生良好互动，并注意新建筑与老建筑的协调与对话关系。同时，注意第五立面的设计。

（2）空间创造

强调室内外空间的创造以及相互关系的良好结合，观赏者不仅能欣赏馆内收藏品，同时也能欣赏户外宜人的景色。各类展示空间宜有视觉上的流通，使彼此有密切的联系。

（3）功能组织

组织好内部功能分区和流线组织，观赏流线宜清晰易辨。

（4）建构表达

①建筑表皮材料肌理表达。

②改造、改扩建设计：需有仓库原建筑结构体系、空间体系、建筑材料分析以及改扩建后的空间、表皮材料的分析和表达。

2. 城市规划要求

（1）建筑密度不得大于 25%，绿地率不得小于 50%。

（2）室外停车：室外临时机动车位 15 个（3mx 6m/个）。

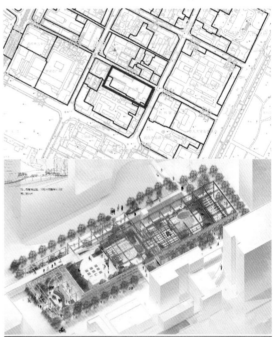

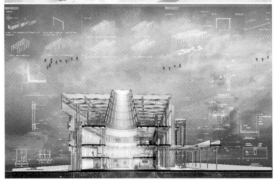

作业范例：彭丹，冉静雅

课题一：后浪时代的大学空间

Topic 1: University Space in the Post Wave Era

● 模型图　　Model diagram

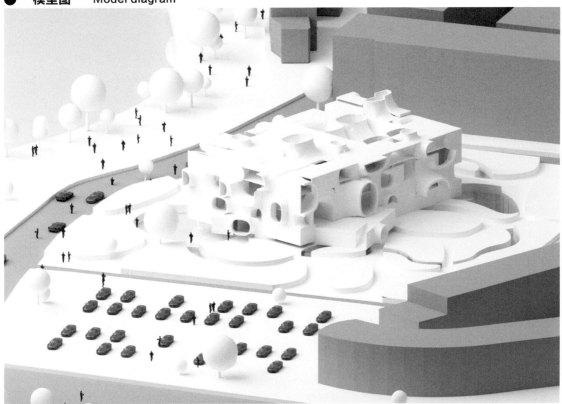

● 透视图　　Perspective view

设计题目： 情绪博物馆 ── 我与黑猫
获奖名称： 2020 年湖南省可持续建筑设计竞赛一等奖
指导老师： 李旭、许昊皓
学　　生： 和译

Design topic: Emotions Museum—I and the Black Cat

Award: The First Prize of Hunan Sustainable Building Design Competition in 2020

Instructors: Li Xu, Xu Haohao

Student: He Yi

● **设计说明**

《情绪博物馆》包括情绪体验与情绪疗愈两部分。前者对情绪进行空间转译，帮助学生了解和应对情绪；后者结合场地的动物与自然，创造疗愈空间。未来大学应当更关注学生的精神层面，营造一个安全、舒适、健康的成长环境。根据场地温度和湿度分析，为获得更好的采光，体块顺应场地进行扭转。我们利用网格对单元体进行切割，将极小曲面单元组合生成情绪空间。同时结合建筑空间风模拟，确定建筑立面洞口和开窗。我们希望把后湖的风和光引入建筑，让每一个游览者感受到自然。

Design notes

The Emotional Museum consists of two parts: the experience of emotions and the healing of emotions. The former expresses emotions by the design of space, so as to help students understand and cope with emotions; The latter combines the animals and nature of the site to create a healing space. In the future, universities should pay more attention to students' spiritual world and create a safe, comfortable and healthy environment for their growth. According to the analysis of the site's temperature and humidity, in order to obtain better lighting, the volume is adjusted in accordance with the site. We use the grid to cut the unit body, and combine the minimal curved elements to generate the emotion space. At the same time, we combine the wind simulation of the building space to determine the entrance and the window of the building. We hope to bring the wind and light from the back lake into the building, so that every visitor can feel the nature here.

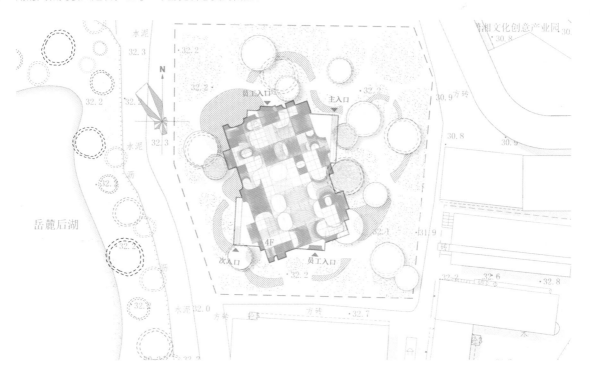

● **空间语法：选型与优化** Spatial grammar: selection and optimization

一只猫的空间？ + 场地的自然语义 + 未来大学科技感 = 极小曲面？

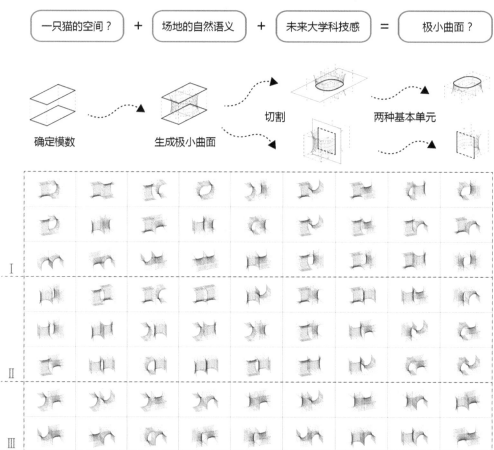

确定模数　　　　　生成极小曲面　　　切割　　　　　两种基本单元

● **建筑体块生成** Building block generation

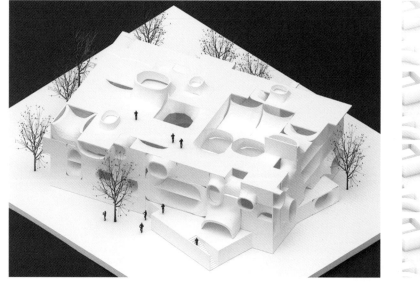

● **情绪体验：与黑猫一同生活**　Emotional experience: living with the black cat

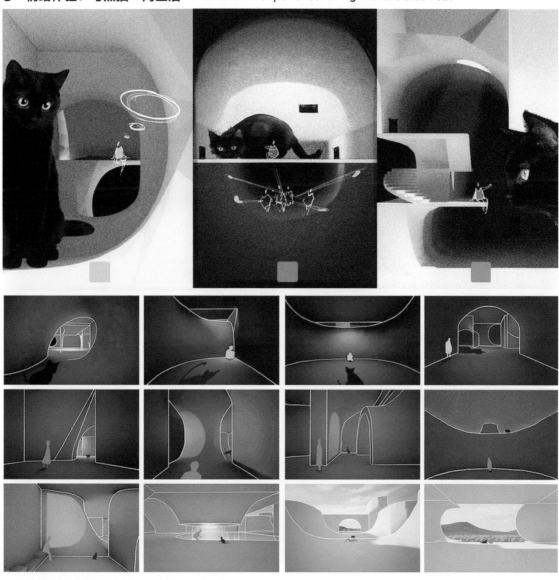

● **立面与剖面图**　Elevation and section

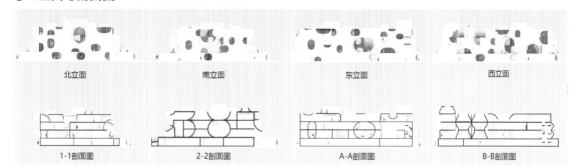

北立面　　　　　　南立面　　　　　　东立面　　　　　　西立面

1-1剖面图　　　　2-2剖面图　　　　A-A剖面图　　　　B-B剖面图

设计题目：积木公寓 —— 未来宿舍模式设计
指导老师：蒋甦琦
学　　生：魏夏清、胡菲

Design topic: Building Block Apartment—Design of Dormitory Mode in the Future

Instructor: Jiang Suqi

Students: Wei Xiaqing, Hu Fei

● 设计说明

在高等教育普及的未来，由于入学人数增加，大学面临扩张，用地短缺，现存宿舍难以满足学生需求。这时，建设快捷而用地不限的临时宿舍，便成为解决问题的关键。教育信息化带来无限便利的同时，也让学生的现实感知弱化，信息接收片面。这时，面对面的交流以及周围事物的信息刺激就变得尤为重要。因此，具有丰富要素和多重体验感的临时宿舍成为设计的主题。在这种宿舍模式下，移动的单元打破了空间的局限性，宿舍模块的个性化定制为自我教育的完善提供了可能。

Design notes

In the future, higher education will be popularized, and universities are facing the problems of expansion and the shortage of land. Therefore, the existing dormitory can't meet the needs of students. Then, the key to solve the problem is to build a fast and unlimited temporary dormitory. While educational informatization brings infinite convenience, it also weakens students' perception of reality and makes information reception one-sided. At this time, face-to-face communication and information stimulation of surrounding things become particularly important. Therefore, the temporary dormitory with rich elements and multiple experiences becomes the theme of the design. In this dormitory mode, the mobile unit breaks the limitation of space; and the personalized customization of dormitory module makes it possible to improve self-education.

● 定制化模块

Customized module

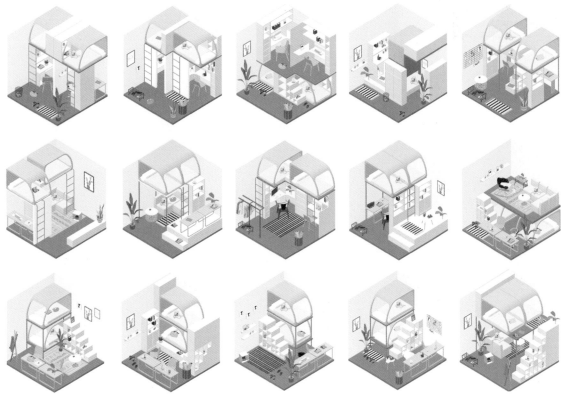

● 体块分析　Block analysis

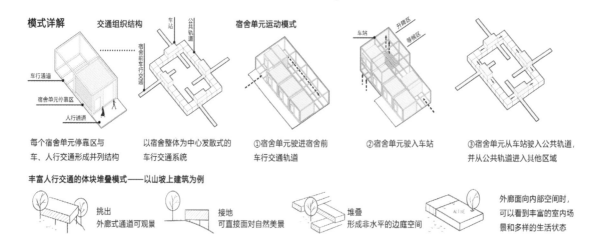

模式详解　　交通组织结构　　　　　宿舍单元运动模式

车行通道

宿舍单元停靠区

人行通道

每个宿舍单元停靠区与
车、人行交通形成并列结构

以宿舍整体为中心发散式的
车行交通系统

①宿舍单元驶进宿舍前
车行交通轨道

②宿舍单元驶入车站

③宿舍单元从车站驶入公共轨道，
并从公共轨道进入其他区域

丰富人行交通的体块堆叠模式——以山坡上建筑为例

挑出
外廊式通道可观景

接地
可直接面对自然美景

堆叠
形成非水平的边庭空间

外廊面向内部空间时，
可以看到丰富的室内场
景和多样的生活状态

● 鸟瞰图　Aerial view

局部透视图　Partial perspective

积木公寓使用指南　Guide to building block

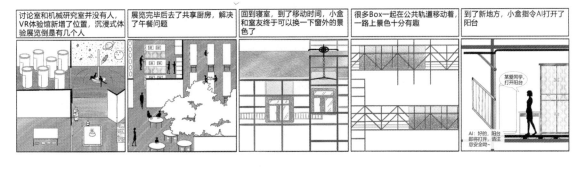

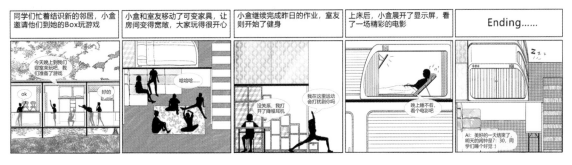

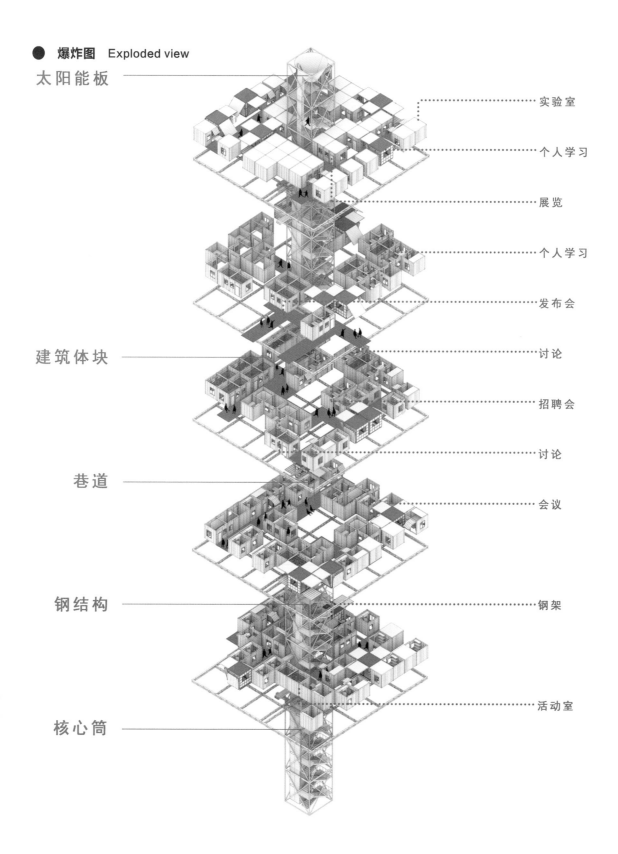

● **爆炸图** Exploded view

太阳能板

实验室

个人学习

展览

个人学习

发布会

建筑体块

讨论

招聘会

讨论

巷道

会议

钢结构

钢架

核心筒

活动室

设计题目："寄生"大学
指导老师：蒋甦琦、张蔚
学　　生：周玥、阿侯伊木、余维、杨明智

Design topic: "Parasitic" University

Instructors: Jiang Suqi, Zhang Wei

Students: Zhou Yue, A Houyimu, Yu Wei, Yang Mingzhi

● 设计说明

在未来机器将取代大量劳动力，底层人民落后于技术进步的更新换代。且教育资源分配不公，底层人民得到同等的教育资源可谓难上加难。

本设计——"寄生"大学致力于创造适用于底层人民学习模式的自发性生长的大学，解决弱势群体的生存、就业与再学习问题。其寄生于三所大学之间的城中村，利用高校教育资源和城中村的市场活力，促进学科知识、产出内容与市场关系的相互融合与相互影响。

Design notes

In the future, machines will replace a large number of labor force, and people in the bottom strata will lag behind the technological progress. Moreover, the unfair distribution of educational resources makes it more difficult for the people at the bottom of the society to get the same educational resources.

This design--"'parasitic'university" is committed to creating spontaneously growing universities that are suitable for the learning mode of the people at the bottom, so as to solve the survival, employment and relearning problems of the vulnerable groups. It parasitizes in the village in the city between the three universities, and makes use of the educational resources of universities and the market vitality of the village in the city to promote and affect each other among the subject knowledge, output content and market relationship.

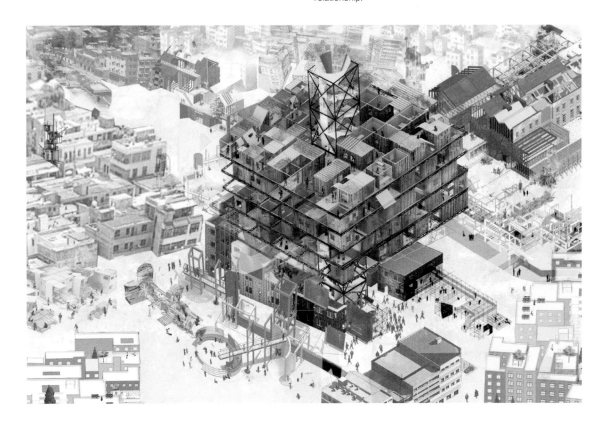

● 概念来源 Concept source

学习模式 Learning model

● 基地选址 Site selection　业态分析 Format analysis

● 日照分析 Sunshine analysis

这些学习者将在城中村内插入廉价的单体作为学习体块，随着人数逐渐增多，逐渐发展为"寄生"大学。

若采用传统垂直叠加的加建方式，原建筑采光会被严重影响，且新建筑单体间相互独立，缺乏交流。

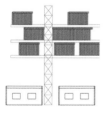

将加建部分架起在老建筑上方，原建筑采光将不受影响，且新建筑单体间彼此连接，增加开放性。

日照控制

影响因子
太阳轨迹 + 建筑分布
ladybug分析

分析老建筑日照劣点，插入核心筒

影响因子
赤纬角、纬度、时角
太阳高度角 + 太阳方位角

分析新老建筑之间的光域范围，使其采光不互相干扰

影响因子
太阳轨迹 + 建筑密度
改版"遗传"算法

分析新建筑之间的单体日照，避免暗房出现

增加开放性

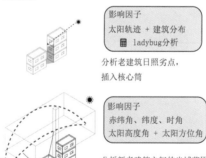

将单体表皮进行改造，增加其开放性

沿核心筒增加研讨、宣讲会等交流空间

● 生长模式　Growth pattern

STEP 1
ladybug 计算原场地日照时长，在日照劣点处插入核心筒，改善老建筑采光。

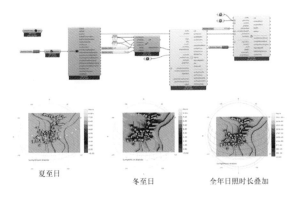

夏至日　　　冬至日　　　全年日照时长叠加

STEP 2
每个核心筒结构可承受最大跨度为 36m，新建筑高度取 5 层计算，公式求太阳方位，保证新老建筑采光满足长沙住宅南向窗日照时长最低标准。

两组新建筑南北向距离为x 两组新建筑东西向距离为y 新建筑高度为H，取15m，新建筑与老建筑间隔距离为h取7.3m

周围新建筑建造后南不影响原建筑南向窗11时 -1时的日照

南北向：x+L = (h+H) cosAw/ tanAs
东西向：y+D = (h+H) sinAw/ tanAs
故 x≥30.5m y≥7.8m

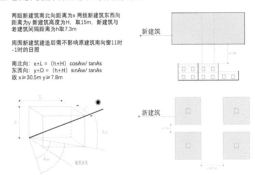

STEP 3
新建筑若在限定范围内满铺，则导致建筑内存在大量暗房。故基于遗传算法的理论，我们把建筑密度和总建筑日照时长作为评判标准，经过多次迭代，让二者达到帕累托最优（即在交换的过程中双方满足达到最大化）。

STEP 4
在计算模拟生成的模型内，取其每层平面的无采光空间最短连接线作为水平交通，并与核心筒相连，共同组成交通体系。

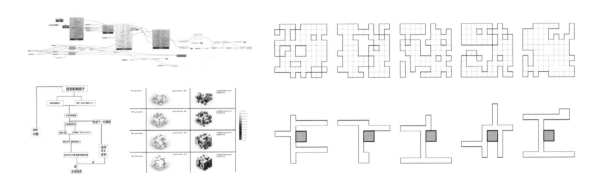

● 场景节点　Scene node

● **局部场景** Local scene

● **总平面图** General layout

● **夜景效果图** Nightscape presentation

设计题目： 边幕 —— 学习共享综合体
指导老师： 吕瑞杰
学　　生： 张晨晖

4

Design topic: Side Curtain—Learning Sharing Complex
Instructor: Lv Ruijie
Student: Zhang Chenhui

● **设计说明**

建筑的架空既对原场地古树进行了保护，同时也解放了地面空间，形成了大片的绿地和室外活动场所，各种有趣的事件都可以在这里发生。

设计将必要的辅助用房和私密空间用墙体进行围合限定，尽可能地将公共区域做到自由、流动、开敞，通过不同建筑体块和微妙的高差来对大的空间进行限定，确定不同的功能分区。

Design notes

The overhead of the building not only protects the old trees in the original site, but also liberates the ground space, forming a large area of green space and places for outdoor activities, where all kinds of interesting events can take place.

The necessary auxiliary room and private space are enclosed and limited by walls so as to make the public area free, flowing and open as far as possible. The large space is limited by different building blocks and subtle height difference to determine different functional zones.

● **概念生成**   **Concept generation**

湖大开放的校园生活	→	往来络绎的学生		粗暴的挡土墙划分		利用透明单纯的半室外建筑空间当作边界，梳理居民和学生及场地的矛盾		建筑空间就像舞台，重新划分了学生和居民的角色，学生成为演员，居民成为观众，角色定位被重置
↓		平静宁和的校园氛围		混乱的道路组织结构				
学生与居民交汇的节点	→	工训中心边界的地理位置	→	居民无意义的穿行	→		→	
↓		具有历史价值的古树		相对封闭的空间模式		利用明净简单的多功能建筑空间当作舞台，激发学生学习的热情		研讨的学生成为主角，起到骨架的作用，成为场地视觉的焦点；自习的学生成为配角，奠定场地的氛围
湖大建筑功能重置利用重要的一步		周边众多的教学楼、院楼		没有活力的场地基调				

居民学生冲突

通过边界重置
二者关系

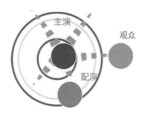

主演辐射整个
建筑舞台

配角营造整个
舞台氛围

1

● **效果图** Design sketch

5

设计题目：凯雪面粉厂改造 —— 现代艺术博物馆设计
指导老师：李旭
学　　生：张嘉晟

Design topic: Kaixue Flour Mill Renovation—Design of Modern Art Museum

Instructor: Li Xu

Student: Zhang Jiasheng

● **设计说明**

场地对社区与城市呈现开放的姿态：用改造的铁轨步道将城市人流引入场地内；在西部和北部界面使建筑空间与城市空间达成一种"媾和关系"；建筑的六部分结构（叙事篇章）界面相互摩擦，使在内部的人清晰地感受到空间的转换；大屋顶体现了一种聚集的姿态和统一的精神。

Design notes

The site is open to the community and the city: the urban pedestrian flow is introduced into the site with the reconstructed railway track; At the interface between the West and the north, a "peace relationship" is reached between architectural space and urban space; The interface of the six part structure (narrative chapter) of the building rubs against each other, so that people inside can clearly feel the transformation of space; The big roof embodies a gathering posture and unified spirit.

● **总平面图**

Master plan

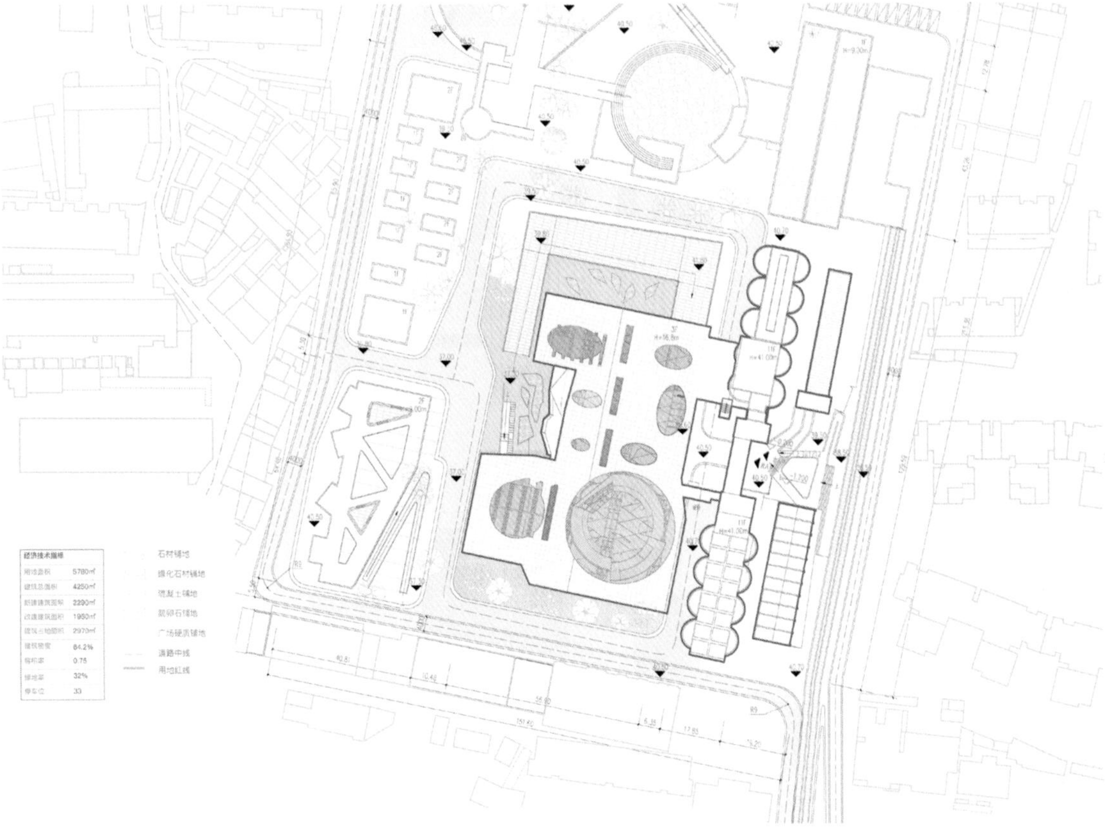

● 叙事文本结构　Narrative text structure

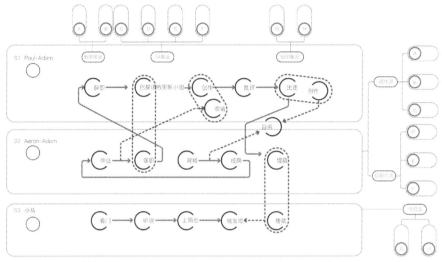

注：S1：A.跟仁·德拉特洛瓦 B.古斯塔夫·崔尔贝 C.皮佳尔·奥古斯特·雷诺阿 D.埃德加·德加 E.克劳德·莫奈 F.爱德华·马奈 G.文森特·梵高 H.保罗·高更
S2：A.勒·柯布西耶 B.密斯·凡德罗 C.马塞尔·布劳耶 D.约瑟·凯奇 E.李迪·詹盖 F.杰夫·沃尔
S3：A.杰夫·昆斯 B.慕国强

● 场地物体提取、编码与拆解　Extraction, coding and dismantling of site

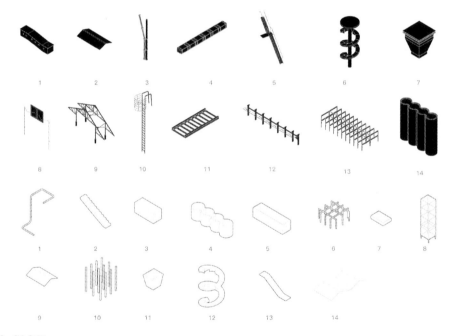

● 空间生成过程　Space generation process

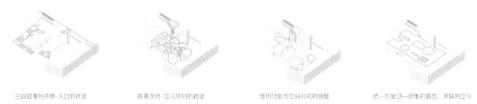

三段叙事的开端-入口的转译　　　叙事序列-空间序列的转译　　　使用功能与空间形式的调整　　　统一的屋顶-聚集的姿态，关联的空间

● 设计策略　Design strategy

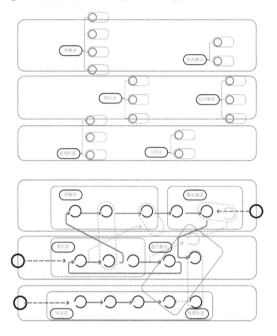

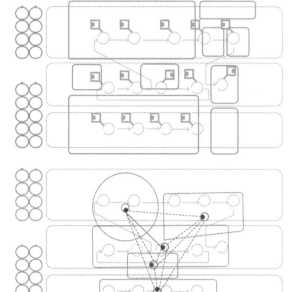

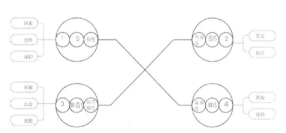

（1）确定展陈作品的六个流派与相应的17位艺术家信息。

（2）将六个流派分别置入文本的三段叙事之中，建立不同流派作品与故事相关情节的关联。

（3）赋予进行类型整理的场地中的物体场所意义，对相应的空间形式进行能指和所指的替换与变形。

（4）使用格雷马斯的二元符号矩阵建立二元对立矩阵，把具体情节置入相应元素中，得到每种元素对应的情节。

● 剖透视　Perspective of dissection

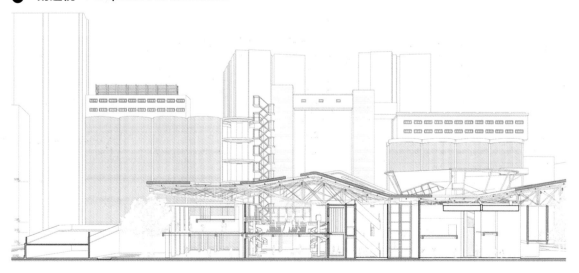

建成效果　Construction effect

设计题目：长沙市工业建筑遗产体系化保护性再利用
　　　　　　—— 潘家坪路 X 号博览馆

指导老师：陈翚

学　　生：陈颂

Design topic: Systematic Protective Reuse of Industrial Architectural Heritage in Changsha

　　　　　　—Expo Hall X, Panjiaping Road

Instructor: Chen Hui

Student: Chen Song

6

● 区位元素提取

面粉厂位于长沙市开福区，周边环境居住区较多，外围将增加办公商务等功能。"严禁烟火"是面粉厂内随处可见的标语，作为工厂的场所记忆提醒着人们这里发生过的历史；长沙，随处可闻"人间烟火"气息，这与本方案的改造方向——面向未来、贴近居民生活的博览建筑主题不谋而合。结合两者，以吸纳"烟火"为目标进行下一步设计。

Location element extraction

The flour mill is located in Kaifu District of Changsha City, where there are many people living in the surrounding area, and office and business functions will be added in the periphery. "No fireworks" is a slogan that can be seen everywhere in the flour factory. As the memory of the factory, it reminds people of the stories here. Changsha is full of the worldly atmosphere, which coincides with the theme of our plan-- the future oriented Expo building, which is close to the residents' life. Combining the two, the next step of design is to absorb the spirit of passion for life.

长株潭城市群
"湘江工业遗产廊道"

长沙市"一江一洲两岸"
工业遗产观光模式

沉浸式

虚拟现实

可互动体验

■ 道路分析

■ 流线分析

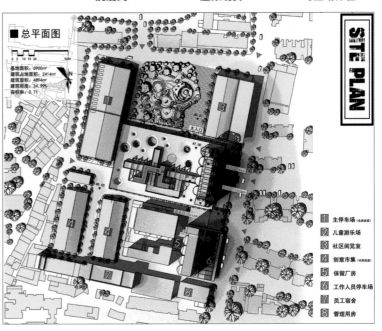
■ 总平面图
SITE PLAN

基地面积：6900㎡
建筑占地面积：2414㎡
建筑面积：4894㎡
建筑高度：34.99%
容积率：0.71

1 主停车场
2 儿童游乐场
3 社区阅览室
4 创意市集
5 保留厂房
6 工作人员停车场
7 员工宿舍
8 管理用房

● 原型转译与重构　Prototype translation and reconstruction

STEP①：交融

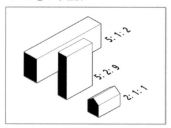

体块来自形态原型1、2、3
图示分别以黄、蓝、红三种颜色代表三个
体块单元

将体块打散
引入形式原型2阵列排架的空间逻辑

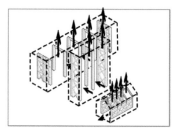

引入形式原型2交叉钢架
以钢架的纹理限定排架单元的空间单位
进而形成空间逻辑

STEP②：攀缘

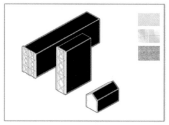

在框架之上生长出相应的材质
这是森林中不同物种的表皮

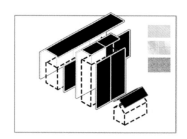

材质与框架剥离
即将自主攀缘与生长

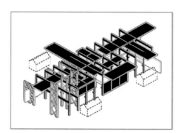

以场地原有厂房的桁架为骨架
新生表皮自由地攀缘生长
实现新旧建构体系的和谐共生

STEP③：重生

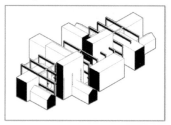

体块单元置入场地原有厂房的桁架系统
组合形成多种空间体验

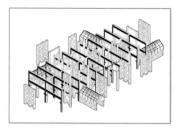

新旧建构体系形成交流
二者的空间逻辑互相穿插交融
二者在交流中得到激活

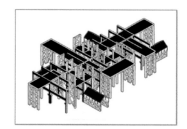

以旧桁架为限定和依托
新"物种"的骨架与表皮肆意生长
元气森林获得新生

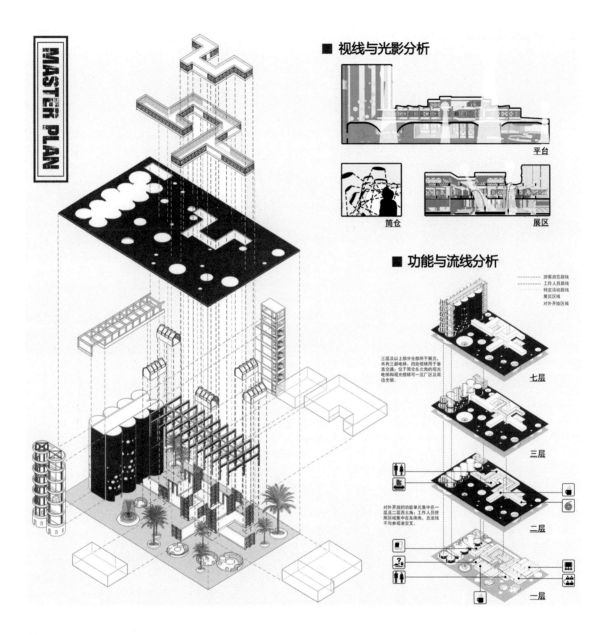

■ 视线与光影分析

平台

筒仓

展区

■ 功能与流线分析

············· 游客游览路线
············· 工作人员路线
············· 特定活动路线
展览区域
对外开放区域

七层

三层及以上部分全部用于展览，共有三部电梯、四处楼梯用于垂直交通；位于筒仓东北角的观光电梯和观光楼梯可一览厂区及周边全貌。

三层

二层

对外开放的功能单元集中在一层及二层西北角，工作人员使用区域集中在东西南角，且流线不与参观者交叉。

一层

● 提取原型　Prototype extraction

形态原型①

浅蓝色涂料粉刷的平顶高层居民楼
建于2010年代
长宽高比约为5:2:9
门窗形成弧形线条，呈横向带状分布

形态原型②

红色坡屋顶多层居民楼
建于1970年代
长宽高比约为2:1:1
门窗呈规则散点状分布

形态原型③

黄灰色涂料粉刷的平屋顶多层居民楼
建于1990年代
长宽高比约为5:1:2
门窗以竖向为主，部分建筑有横向线条装饰

课题三：城市慢生活：社区图文信息中心设计
Topic 3: Urban Slow Life: Design of Community Graphic Information Center

● 鸟瞰图　Aerial view

● 效果图　Design sketch

● 局部场景　Local scene

设计题目： 社区图文信息中心 —— 壁隐

指导老师： 李旭

学　　生： 赵伯轩、张嘉晟

7

Design topic: Community Graphic Information Center—Biyin

Instructor: Li Xu

Students: Zhao Boxuan, Zhang Jiasheng

● **设计说明**

基于安静且富场景感的场地，我们决定在场地引入一片人在其中可居可游、可行可望的"山水"。通过对《天池石壁图》手法的分析和转译，赋予场地一系列意趣丰富的空间游线。本方案用统一的原型与手法在主要阅览空间等营造叙事空间，而后用几条游线串联场地。最后形成的几个空间叙事场景，分别为：穿林问山、俯仰洞天、重岩叠嶂、西临碣石、深崖飞瀑、曲水流觞。

Design notes

Based on the quiet and scene feeling of the site, we decided to introduce a "landscape" in which people can live, swim and hope The analysis and translation of the technique endows the site with a series of interesting spatial tour lines.

In this scheme, a unified prototype and technique are used to create a narrative space in the main reading space, and then several tour lines are used to connect the site. Finally, several spatial narrative scenes are formed, which are as follows: looking down through the linwen mountain, the heavy rocks in the cave, the waterfalls and water fountains on the Jieshi deep cliff in the West.

● **空间路径图解**

Spatial path diagram

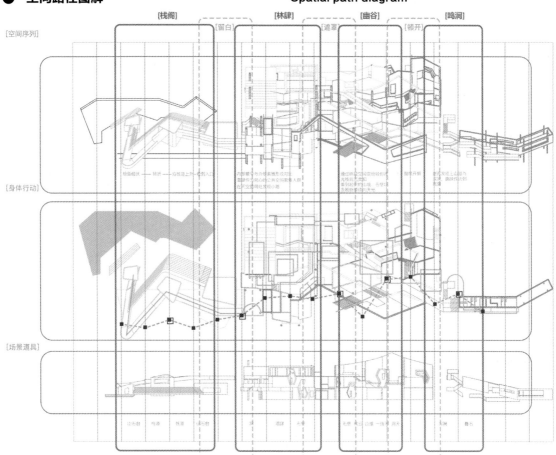

● **模型图** Model diagram

● **场景效果图** Scene rendering

设计题目： 社区图文中心 —— 流动与聚集
指导老师： 李旭
学　　生： 黄钊、祝宸

Design topic: Community Graphic Center—Flow and Gathering

Instructor: Li Xu

Students: Huang Zhao, Zhu Chen

8

● 慢生活行为模式

行走在场地中时，我们发现，社区中居民的生活行为表现出的流动性与聚集性十分引人注目。流动性体现在该社区居民有散步、闲逛的习惯。无论老人还是中青年，都喜欢闲暇时在社区中走动；另外，居民的行为又表现出聚集性，例如聚在一起打牌、打麻将，又如散步看见熟人会驻足聊天，等等。

流动与聚集两个看似矛盾的特性，在这个社区的居民身上得到了完美的融合。所以，我们在社区图文中心的设计上，想到了利用流动与聚集的概念，将这种行为模式应用于阅读氛围的营造。

Slow life behavior pattern

When walking in the field, we find that the mobility and aggregation of residents in the community is very noticeable. Mobility is reflected in the community residents who have the habit of walking and strolling. Both the elderly and young people like to walk around the community in their spare time. On the other hand, the behavior of residents shows aggregation, such as gathering together to play cards and mahjong, and stopping to chat with acquaintances and so on.

The two seemingly contradictory characteristics of mobility and aggregation are perfectly integrated in the residents of this community. Therefore, in the design of community graphic center, we thought of using the concept of mobility and aggregation to apply this behavior mode to the cultivation of reading atmosphere.

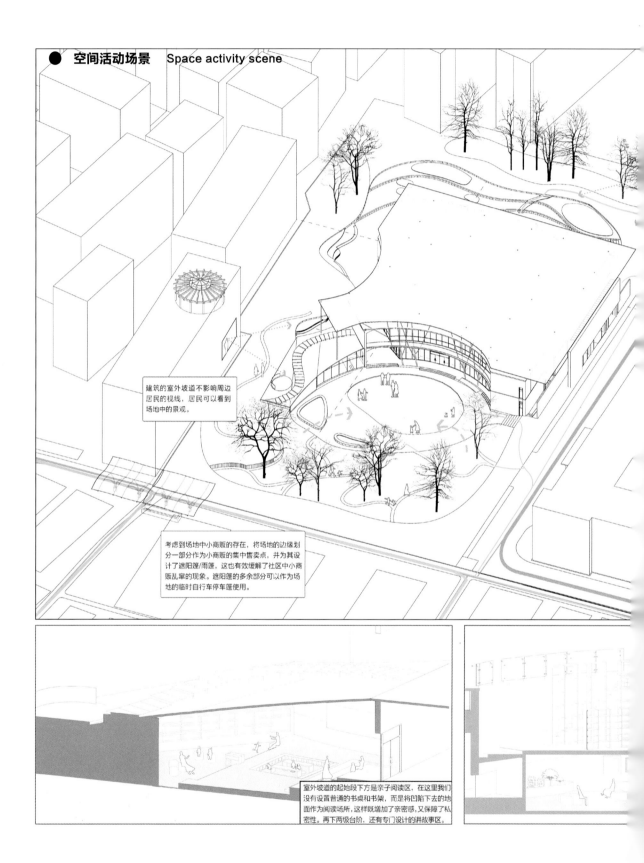

空间活动场景　Space activity scene

建筑的室外坡道不影响周边居民的视线，居民可以看到场地中的景观。

考虑到场地中小商贩的存在，将场地的边缘划分一部分作为小商贩的集中售卖点，并为其设计了遮阳蓬/雨蓬，这也有效缓解了社区中小商贩乱窜的现象。遮阳蓬的多余部分可以作为场地的临时自行车停车蓬使用。

室外坡道的起始段下方是亲子阅读区，在这里我们没有设置普通的书桌和书架，而是将凹陷下去的地面作为阅读场所，这样既增加了亲密感，又保障了私密性。再下两级台阶，还有专门设计的讲故事区。

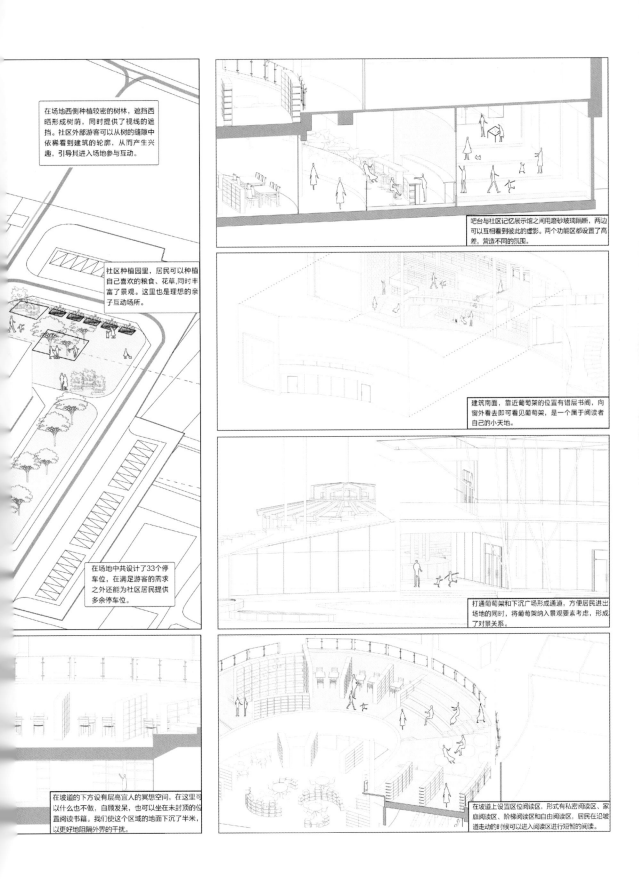

在场地西侧种植较密的树林，遮挡西晒形成树荫，同时提供了视线的遮挡。社区外部游客可以从树的缝隙中依稀看到建筑的轮廓，从而产生兴趣，引导其进入场地参与互动。

社区种植园里，居民可以种植自己喜欢的粮食、花草，同时丰富了景观。这里也是理想的亲子互动场所。

在场地中共设计了33个停车位，在满足游客的需求之外还能为社区居民提供多余停车位。

吧台与社区记忆展示馆之间用磨砂坡璃隔断，两边可以互相看到彼此的虚影。两个功能区都设置了高差，营造不同的氛围。

建筑南面，靠近葡萄架的位置有错层书阁，向窗外看去即可看见葡萄架，是一个属于阅读者自己的小天地。

打通葡萄架和下沉广场形成通道，方便居民进出场地的同时，将葡萄架纳入景观要素考虑，形成了对景关系。

在坡道的下方设有层高宜人的冥想空间，在这里可以什么也不做，自顾发呆，也可以坐在未封顶的位置阅读书籍。我们使这个区域的地面下沉了半米，以更好地阻隔外界的干扰。

在坡道上设置区位阅读区，形式有私密阅读区、家庭阅读区、阶梯阅读区和自由阅读区，居民在沿坡道走动的时候可以进入阅读区进行短暂的阅读。

08:05

心理问题日益成为大学生群体中普遍性的健康问题，繁重的学业压力、人际关系、生存压力使近九成的大学生产生情绪问题，疏导不良情绪、充分感知自我变成一个不容忽视的人生课题。而大学是典型的集体生活，大学生能够在一个空间中独处的机会很少；与此相矛盾的是，多数大学生的社交圈又被禁锢在以寝室为单位的界限中。故为大学生提供与自己相处的空间，与他人相遇的空间，选择教学区与宿舍区之间的场地，有连接学习与生活的意味，更有从学习生活中跳脱出来的另一重意味。

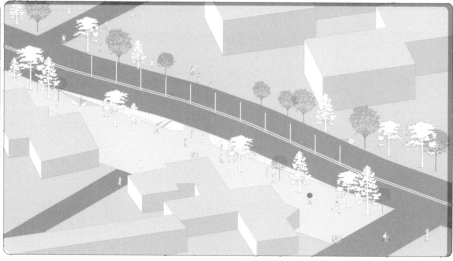

● 方案过程　Program process

从传统天桥形式入手，将功能区放在桥上。压下屋顶的两对角从而形成一条自行车道，但因道路宽度较小无法满足道路坡度限值，从而考虑到将直桥与引桥结合变为环形。

一草

与一草形体生成逻辑相似，将屋顶部分压至地平面使屋顶与一层形成通路，但因入口部分人、车行驶问题将设计引向死胡同。此时产生了一个疑问：地面的路权一定属于机动车吗？

二草

交通解决方案

形体生成逻辑

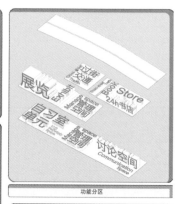

功能分区

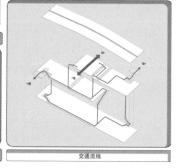

交通流线

106

设计题目： 社区图文中心
指导老师： 张光
学　　生： 周婧依

Design topic: Community Graphics and Text Center

Instructor: Zhang Guang

Student: Zhou Jingyi

● 街道空间重构

心理问题日益成为大学生群体中普遍性的健康问题，繁重的学业压力、人际关系、生存压力使近九成的大学生产生情绪问题，疏导不良情绪、充分感知自我变成一个不容忽视的人生课题。而大学是典型的集体生活，大学生能够在一个空间中独处的机会很少。

与此相矛盾的是，多数大学生的社交圈又被禁锢在以寝室为单位的界限中。故为大学生提供与自己相处的空间、与他人相遇的空间，选择教学区与宿舍区之间的场地，有连接学习与生活的意味，更有从学习生活中跳脱出来的另一重意味。

Street space reconstruction

Psychological problems are increasingly becoming a common health problem among college students. Heavy academic pressure, complicated interpersonal relationship and subsistence pressure make nearly 90% of college students have emotional problems. It is a life topic that can not be ignored to dredge bad emotions and fully perceive themselves. While life in universities is a typical collective life, and there are few opportunities for college students to be alone.

Paradoxically, the social circle of most college students is confined in the boundary of dormitory. Therefore, to provide space for college students to get along with themselves and meet others, and to choose the space between the teaching area and the dormitory area, has the meaning of connecting learning and life, and has another meaning of jumping out of learning and life.

总平面图 1：2000

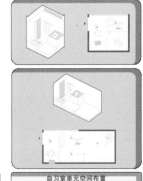

自习室单元空间布置

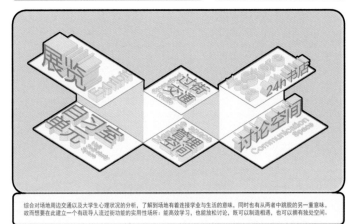

综合对场地周边交通以及大学生心理状况的分析，了解到场地有着连接学业与生活的意味，同时也有从两者中跳脱的另一重意味，故而想要在此建立一个有疏导人流经过街功能的实用性场所：能高效学习，也能放松讨论，既可以制造相遇，也可以拥有独处空间。

App线上预约制

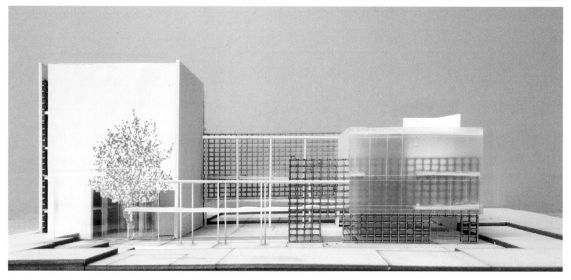

设计题目：校园图文信息中心 —— 势差

指导老师：张蔚

学　　生：孙智霖

10

Design topic: Campus Graphic and Text Information Center—Potential Difference

Instructor: Zhang Wei

Student: Sun Zhilin

● **总平面图**　**General plan**

● 体块生成　Block generation

1. 基地现状：场地内部存在较低高差且场地周边存在保护树木，其北侧校园道路在上下学时间车流量较大。
* 场地高差被调整为18m（以绝对标高49.7m为起始高度计算）。

2. 按右侧蜿蜒脚步建立控制线，确定建筑主要体量位置，并下沉营造汇览广场。

3. 下沉空间向周围蔓延，整合与场地的关系，图书馆东侧绿化延伸使场地内绿地延续，延续东楼入口通路，激活图书馆西侧庭院。建筑体块对控制线位置进行填充。

4. 建筑中部置入庭院，并将外侧建筑高度适当降低，对人做出欢迎的姿态，与现有图书馆对应的同时营造出汇览中心。

5. 建筑东侧高度下降至新图书馆裙楼高度，在东侧主干道立面与原有图书馆高度呼应，与新馆形成错落层次。庭院入口处建筑体块简化为室外流线，进一步加强庭院通透性，同时架空入口处比领体块，与校园主干道沟通。

6. 两侧体块置入与中部庭院中轴对称的小庭院（个体意识），加强与现图书馆的轴线关系。

● 概念与形式　Concept and form

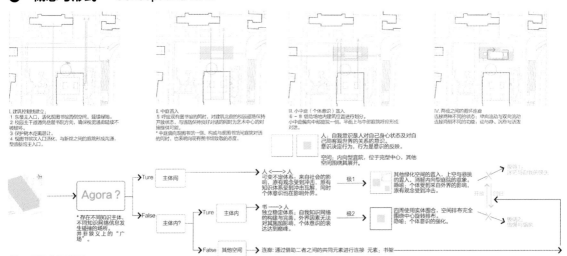

I. 建筑控制线建立：
1. 东楼主入口，活化即图书馆西侧空间，延续绿地。
2. 校园主干道通向总体图书馆方向，确保视觉通廊延续不被破坏。
3. 保护场地内乔木适需退让。
4. 现图书馆入口活化，与新馆之间的庭院简形成沟通，整理新馆主入口。

II. 中庭置入：
5. 呼应原有图书馆的同时，对建筑北侧的校园道路保持开放的姿态，与道路保持较好对话的同时为艺术中心的对接提供可能。
* 中庭偏向图书馆一侧，构成与现图书馆间庭院对话。
中庭偏向现图书馆一侧的同时，也表明了新馆对现有图书馆致敬的态度。

III. 小中庭（个体意识）置入：
6～8 借助场地与建筑位置进行划分。
小中庭偏向现图书馆一侧，平面上与中部庭院相呼应形成对话。

人：自我意识是人对自己身心状态及对自己同客观世界的关系的意识。意识决定行为，行为是意识的反映。

空间：内向型庭院，位于完整中心，其他空间围绕其展开。

IV. 两级之间的循环连接
连接两种不同的状态：单向流动与双向流动。连接提别不同的功能：动与静、沉隐与活泼。

Agora？
* 存在不同知识主体，不同知识网络信息发生碰撞的场所。并非狭义上的"广场"。

Ture → 主体间
False → 主体内？ → Ture → 主体内
　　　　　　　　 → False → 其他空间

人←→人
可变化关系；素质社会的影响，原有观念受到冲击；原有知识体系与冲击瓦解，同时个体意识也在影响外界

书——人
独立稳定体系；自我知识网络的构建与完善，外界因素无法对其施加影响，个体意识的表达达到巅峰。

极1 → 其他绿化空间的置入，上穿与悬挑的置入，消解内向型庭院的盛堂，隐喻：个体受到来自外界的影响，原有观念受到冲击。

极2 → 四面使用实体围合，空间排布完全围绕中心旋转排布。隐喻：个体意识的强化。

连廊：通过借助二者之间的共同元素进行连接 元素：书架

● 南立面图　South elevation

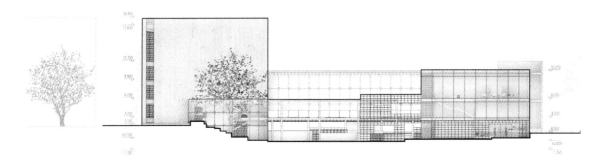

● 首层平面图　First floor plan

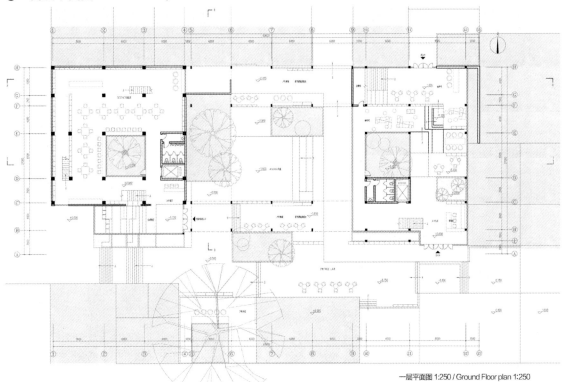

一层平面图 1:250 / Ground Floor plan 1:250

● 剖面构造　Section structure

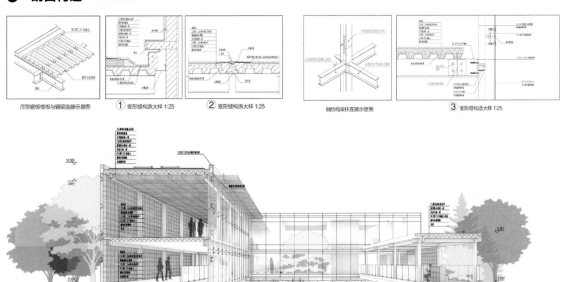

压型钢板楼板与钢梁连接示意图　　① 变形缝构造大样 1:25　　② 变形缝构造大样 1:25　　钢结构梁柱连接示意图　　③ 变形缝构造大样 1:25

● 分层轴测图　Layered axonometric drawing

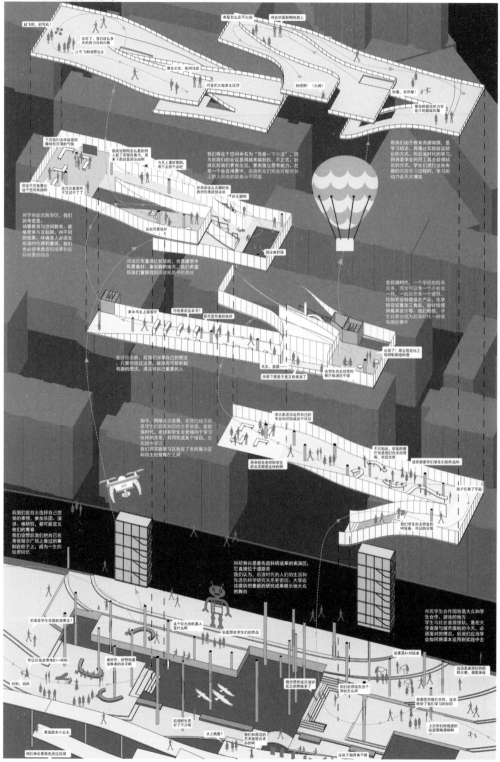

11

设计题目：浮于事
获奖名称：2020 年谷雨杯全国大学生可持续建筑设计竞赛一等奖
指导老师：彭智谋
学　　生：宋智明、王昊、方万俊、庞艳

Design topic: Floating on Things

Award: First Prize of 2020 Guyu Cup National College Students' Sustainable Architectural Design Competition

Instructor: Peng Zhimou

Students: Song Zhiming, Wang Hao, Fang Wanjun, Pang Yan

● "浮于事"

"人浮于事"原指僧多粥少、人多事少。如果把"事"引申为校园中的各种事件，那"人浮于事"则指：后浪学子越来越多，但校园里美好的事件却又不断减少。

而这个建筑"浮于事"，既是在指建筑体量漂浮在各种事件之上，也是想说：没有事件，我们的校园将无处落脚。我们希望这个建筑成为一个校园事件的发生装置，并以开放的象牙塔的形象展示在公众面前。

"Floating on things"

"Have more people to deal with things than are needed" originally refers to overstaffing. If "things" are extended to various events in the campus, then the phrase means that more and more younger-generation students are appearing, but the good events in the campus are decreasing.

The building "Floating on Things" means that the building is floating on various events, but also means that without events, our campus will have no place to settle. We hope that the building will become a device for campus events and be displayed in the public as an open ivory tower.

灯光文化节

学生市民互动展示区

商量一下小道和泳池

岳麓山渔场	严重污染城中村	锦绣潇湘文化园	地铁修建、都乐街拆除	后湖区
1958 · 1958 - 2008 · 2008 - 2010	2010 - 2015	2010 - 2015	2015 - 2020	

湘江西岸散居的10多个渔民组成了岳麓山渔场。此时后湖边有渔民们捕鱼的事件发生

锦绣潇湘文化创业产业园建成，后湖区域成为展示、售卖文创手工作品等艺术性产品的聚集地。大量文创相关事件在此发生

"堕落街"拆除，在这里形成了天马夜市，也就是"多乐街"。成了学生们聚集、吃饭的地方

因为地铁建设，多乐街被拆，即天马夜市消失。场地变成事件发生的游乐场

架空空间

体验互动式教学空间

● 空间场景　Space scene

设计题目：IFL 潮汐食学园
获奖名称：2020 年谷雨杯全国大学生可持续建筑设计竞赛特等奖
指导老师：罗苈、严湘琦
学　　生：刘深圳、武文忻、李杼欣、高原

Design topic: IFL Tidal Dining and Learning Park

Award: Special prize of 2020 Guyu Cup National College Students' Sustainable Architectural Design Competition

Instructors: Luo Jin, Yan Xiangqi

Students: Liu Shenzhen，Wu Wenxin, Li Zhuxin, Gao Yuan

● 未来属性场所重构：非正式学习乐园　　Reconstruction of future place: informal learning park

非正式学习是包括信息和内容在内的一切事物，如会议、书籍、网站等，或者是非正式的人与人的交流，例如交谈、讨论、会议等。我们选取了宿舍园区的一个食堂，进行方案设计。根据食堂已有的特点：食学合一、两方潮汐周期性。同时，基于互联网时代下的非正式学习"输入—转化—输出"的学习模式，结合大数据信息共享与反馈机制连接虚拟空间与现实空间，实现人机交互，并重构"信息自动化"食堂。通过可适性单元模块与移动组合模式实现上述模式中"转换"过程的主动性，让未来的大学生自由自在地进行知识内化与人际交往。

Informal learning refers to all things including information and content, such as meetings, books, websites, etc., or informal interpersonal communication, such as conversation, discussion, meeting, etc. In our existing campus environment, there are already such places. Therefore, we selected a canteen in the dormitory park for the scheme design. According to the existing characteristics of the canteen: the unity of eating and learning, and the periodicity of tides on both sides. At the same time, based on the "input, transformation, output" learning mode of informal learning in the Internet era, combined with the big data information sharing and feedback mechanism, the virtual space and real space are connected to realize human-computer interaction and reconstruct the "information automation" canteen. Through the adaptive unit module and mobile combination mode, the initiative of the "transformation" process in the above mode can be realized, so that future college students can freely enjoy the internalization of knowledge and human interaction.

● 可适性模组　Adaptability module

私人剧院
Private Theater

重新成牌
Enrollment

讨论时间
Discussion and
coffee

艺术传结
Art Media

VR沉浸
VR Immersion

回转寿司
Conveyer Belt
sushi

自助桌游
智RPG

艺术体验
Artistic Immersion

社团聚会
Club Activity

微型展览
Diminutive Exhibi-
tion

学术研讨
Academic Study

学习领社区
Small Learning Com-
munity

成果汇报
Achievement
Report

自助专教
Professional
Clasroom

● 食学空间重构策略
Reconstruction strategy of dining and learning

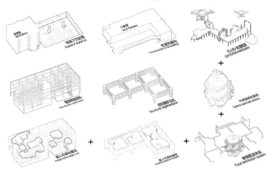

● 食堂筒分区
Canteen barrel partition

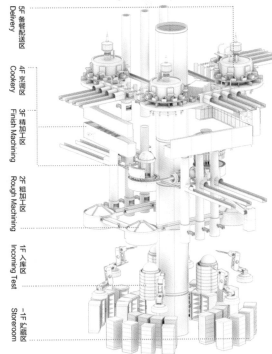

5F 备餐配送区
Delivery

4F 烹调区
Cookery

3F 精加工区
Finish Machining

2F 粗加工区
Rough Machining

1F 入库区
Incoming Test

-1F 贮藏区
Storeroom

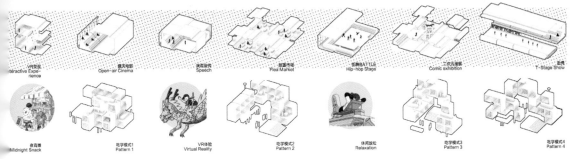

VR交互 Interactive Experience | 露天电影 Open-air Cinema | 演说论坛 Speech | 创意市集 Flea Market | 街舞BATTLE Hip-hop Stage | 二次元漫展 Comic exhibition | T台秀 T-Stage Show

夜宵摊 Midnight Snack | 范学模式1 Pattern 1 | VR体验 Virtual Reality | 范学模式2 Pattern 2 | 休闲放松 Relaxation | 范学模式3 Pattern 3 | 范学模式4 Pattern 4

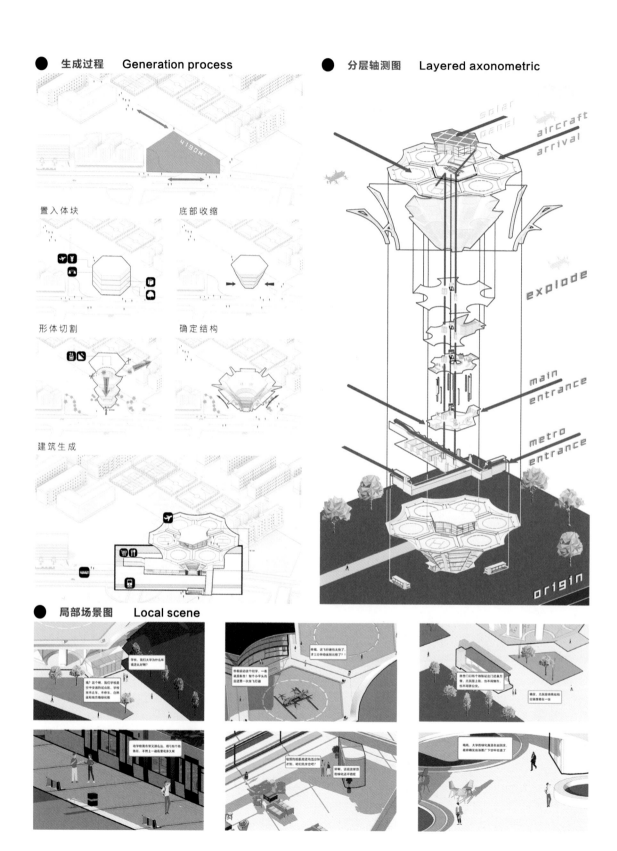

● 生成过程　Generation process

置入体块

底部收缩

形体切割

确定结构

建筑生成

● 分层轴测图　Layered axonometric

solar panel

aircraft arrival

explode

main entrance

metro entrance

origin

● 局部场景图　Local scene

13

设计题目： 校际穿梭
获奖名称： 2020 年谷雨杯全国大学生可持续建筑设计竞赛二等奖
指导老师： 彭智谋、罗苨
学　　生： 李卓恒、赵振宇、骆炳仁

Design topic: Inter School Shuttle

Award: Second prize of 2020 Guyu Cup National College Students' Sustainable Architectural Design Competition

Instructors: Peng Zhimou, Luo Peng

Students: Li Zhuoheng, Zhao Zhenyu, Luo Bingren

● **设计说明**

本次设计，希望通过新技术的应用，改变现有的交通生活方式，从解决大学校区间的联系问题入手，扭转现有技术带来的过于膨胀的地面交通，消解技术的异化影响，让孤岛变为绿洲，解放校园的地面空间，进而一步步影响整个社会，让地面空间回归生态和谐。

Design notes

In this design, we hope to change the existing way of transportation and life through the application of new technology. Starting from solving the problem of connection between university campuses, we hope to reverse the over expansion of ground transportation brought by existing technology, eliminate the influence of technology alienation, turn isolated island into oasis, liberate the ground space of campus, and further affect the whole society step by step, restore the ecological harmony of the ground space.

基于we
太阳辐

ANNUAL INCIDENT SOLAR RADIATION AT 132.0°

DB 电能做动力，比燃烧引擎的噪声更小，同时起降台在距离地30m的高度，基本不会对地面产生噪声影响。

DB = 标准直升机   or   中型卡车
      1/32   1/2

电动垂直起降，可以240km/h以上的速度在305~610m的高度巡航。

¥ 现阶段成本已能低至每英里2美元，在未来这个成本还可以更低。

空中交通对地面的解放使得地面不再需要那么多硬质铺装，绿地的大大增加更容易设计雨水花园，使得海绵城市的理念更容易实现。

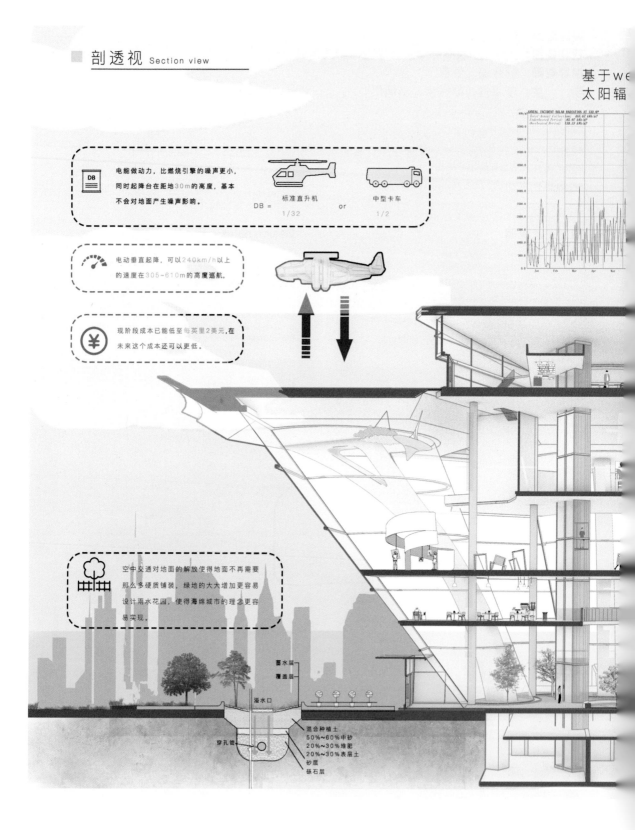

蓄水层
覆盖层
溢水口
穿孔管
混合种植土
50%~60%中砂
20%~30%堆肥
20%~30%表层土
砂层
砾石层

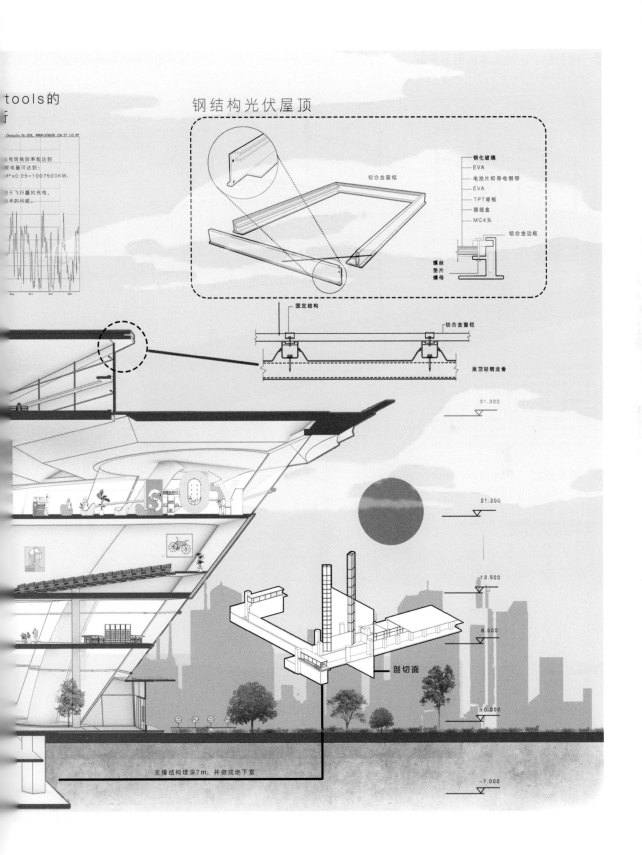

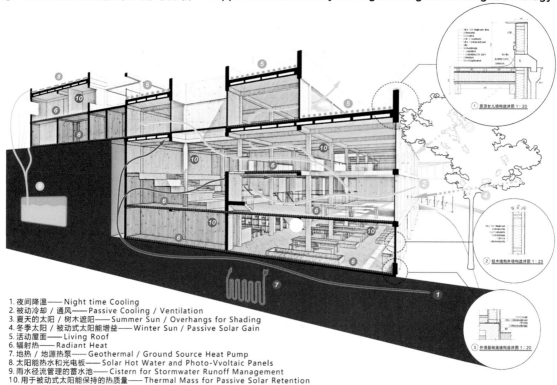

1. 夜间降温——Night time Cooling
2. 被动冷却／通风——Passive Cooling / Ventilation
3. 夏天的太阳／树木遮阳——Summer Sun / Overhangs for Shading
4. 冬季太阳／被动式太阳能增益——Winter Sun / Passive Solar Gain
5. 活动屋面——Living Roof
6. 辐射热——Radiant Heat
7. 地热／地源热泵——Geothermal / Ground Source Heat Pump
8. 太阳能热水和光电板——Solar Hot Water and Photo-Vvoltaic Panels
9. 雨水径流管理的蓄水池——Cistern for Stormwater Runoff Management
10. 用于被动式太阳能保持的热质量——Thermal Mass for Passive Solar Retention

● 场景概念　Scene concept

设计题目: 未来大学 —— 无边界·驿站
获奖名称: 2020 年谷雨杯全国大学生可持续建筑设计竞赛二等奖
指导老师: 许昊皓、陈翚
学　　生: 陆禹名、章纪伟、王文爱、卫静怡

14

Design topic: University of the Future—Borderless Station

Award: Second prize of 2020 Guyu Cup National College Students' Sustainable Architectural Design Competition

Instructors: Xu Haohao, Chen Hui

Students: Lu Yuming，Zhang Jiwei，Wang Wen'ai，Wei Jingyi

● **设计理念**

此次设计主题为"无边界·驿站"，即面对数字化科技高速发展、人的价值逐渐异化的危机时代，利用科技打破知识壁垒，同时将教育本质回归到"人"本身，创造一个集无边界学习、交流、参观游览于一体的学习住宿游学驿站，开启体验式游学新思路。设计从自然因素入手，提取采光、通风数据集，从人流模拟优化平面组织，从场地高差以及人的视线入手生成剖面，并辅助以网络小程序实现驿站与驿站之间的互通有无。

Design notes

The theme of this design is "Borderless Station". It aims to break the knowledge barrier by using science and technology in the face of the crisis era of rapid development of digital technology and gradual alienation of human value. At the same time, it will return the essence of education to "human" itself, create a learning, accommodation and study post with no boundary learning, communication and sightseeing, and open a new idea of experiential study tour. The design starts with the natural factors, extracts the daylighting and ventilation data sets, optimizes the plane organization from the human flow simulation, and generates the section from the site height difference and human line of sight. And auxiliary network small programs are applied to realize the communication between various post stations.

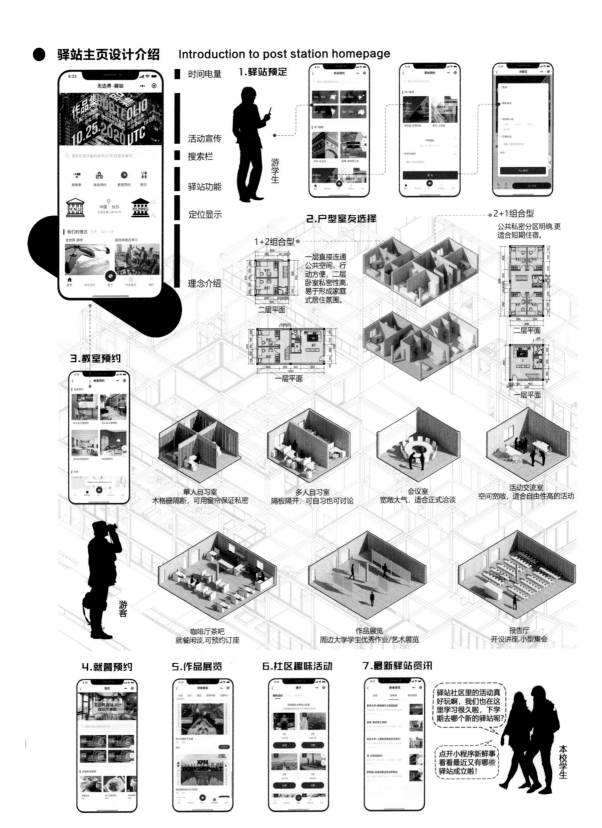

● 驿站主页设计介绍　**Introduction to post station homepage**

时间电量

1.驿站预定

活动宣传

搜索栏

驿站功能

定位显示

理念介绍

游学生

2.户型室友选择

2+1组合型
公共私密分区明确,更
适合短期住宿。

1+2组合型
一层直接连通
公共空间,行
动方便,二层
卧室私密性高,
易于形成家庭
式居住氛围。

二层平面

一层平面

二层平面

一层平面

3.教室预约

单人自习室
木格栅隔断,可用窗帘保证私密

多人自习室
隔板隔开,可自习也可讨论

会议室
宽敞大气,适合正式洽谈

活动交流室
空间宽敞,适合自由性高的活动

游客

咖啡厅茶吧
就餐闲谈,可预约订座

作品展览
周边大学学生优秀作业/艺术展览

报告厅
开设讲座,小型集会

4.就餐预约　　**5.作品展览**　　**6.社区趣味活动**　　**7.最新驿站资讯**

驿站社区里的活动真
好玩啊,我们也在这
里学习很久啦,下学
期去哪个新的驿站呢?

点开小程序新鲜事
看看最近又有哪些
驿站成立啦!

本校学生

● **未来，全球驿站时代** Future, an era of global post station

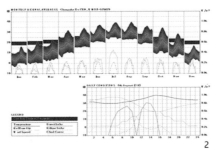

1

2

3

1.搜索全球大学
2.提取地理物理数据
3.人群城市形态模拟
4.建成驿站
5.全球交流增加

4

5

● **场景节点效果图** Scene node rendering

● **场地分析** Site analysis

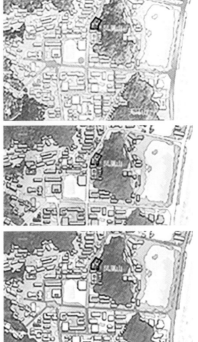

课题五：捷克 Liberec 博物馆改扩建设计
Topic 5: Reconstruction and Expansion Design of Liberec Museum

● **效果图** Design sketch

设计题目： 捷克 Liberec 博物馆改扩建设计
获奖名称： 2017 年中国建筑院校境外交流作业展三等奖
指导老师： 罗荩、蒋甦琦
学　　生： 彭斯佳、张婷

15

Design Title: Reconstruction and Expansion Design of Liberec Museum in Czech Republic

Award: Third prize of 2017 Overseas Exchange Operation Exhibition of Chinese Architecture Colleges

Instructor: Luo Jin, Jiang Suqi

Students: Peng Sijia, Zhang Ting

● **视线游戏**

博物馆改扩建的主要概念是基于新旧博物馆之间的关系，即以旧博物馆为新的展品，以新博物馆为旧的舞台，我们关注的是观看的行为，博物馆内的人们可以在其他博物馆观看其他人的活动，也可以看到建筑的某些部分。当谈到实体时，我们有点受侦探小说的启发，在新建筑中放置了一个由我们的设计师指导的图景，但参观者可以通过多种方式重读。在博物馆里人们将开始他们的参观。

Vision game

The main concept of museum expansion is based on the relationship between the old and new museums, that is, taking the old museum as the new exhibit and the new museum as the old stage. What we pay attention to is the behavior of watching. People in the museum can watch other people's activities in other museums, and can also see some parts of the building. When it comes to substance, it's a bit inspired by detective stories. We put a picture in the new building under the guidance of our own designer, but visitors can reread it in many ways. In the museum, people will begin their visit.

● **形成过程　Formation process**

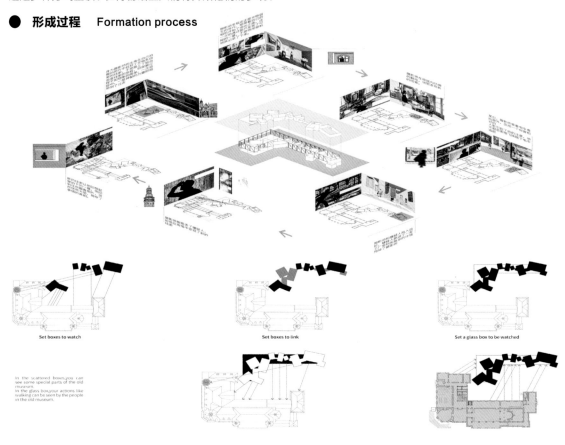

Set boxes to watch

Set boxes to link

Set a glass box to be watched

In the scattered boxes,you can see some special parts of the old museum.
In the glass box,your actions like walking can be seen by the people in the old museum.

entrance hall from the southeast corner   entrance hall from the city new city square

● 效果图　Design sketch

● 整体鸟瞰图　Overall aerial view

设计题目： 城市活力再生 —— 斯柯达工厂城市设计
指导老师： 陈翚
学　　生： 董文涵

16

Design topic: Regeneration of Urban Vitality—Skoda Factory Urban Design

Instructor: Chen Hui

Student: Dong Wenhan

● 项目介绍

本次城市设计基地位于欧洲中部国家捷克的工业城市 —— 姆拉达·博莱斯拉夫。姆拉达·博莱斯拉夫是捷克最富裕的城市之一。斯柯达汽车公司的总部工厂就在这个城市。基地位于斯柯达工厂处。我们认为其主要存在两个问题：
（1）交通问题。每天城市中及周围城市许多人来到斯柯达工厂上班，而现有的交通结构较为单一，导致交通拥堵。
（2）该区域缺乏与城市的联系。目前，该区域零散分布着工厂、学校、停车场等功能，与城市分割严重，导致该区域缺乏活力。

Project introduction

The urban design base is located in Mlada Boleslav, an industrial city in the central European country of Czech Republic. Mlada Bleslav is one of the richest cities in the Czech Republic. Skoda's headquarters and factory are in this city. The base is located at Skoda factory. We think that there are two main problems:

(1) Traffic problems. Every day, many people in and around the city come to Skoda factory to work, but the existing traffic structure is relatively simple, leading to traffic congestion.

(2) There is a lack of connection with cities in the region. At present, some factories, schools, parking lots and other functions are scattered in this area, which is seriously separated from the city, leading to the lack of vitality in this area.

● 总图对应剖断面　General drawing section

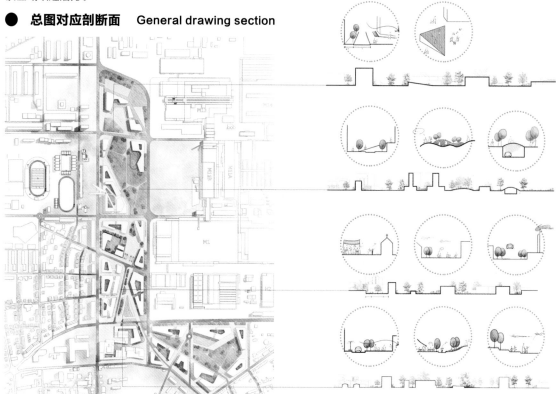

场景效果图　Scene rendering

剖面光影分析　Profile light and shadow analysis

17

设计题目： 光域 —— 捷克城堡博物馆扩建设计
指导老师： 蒋甦琦
学　　生： 焦智恒、刘恬

Design topic: Light Field—Expansion Design of Czech Castle Museum

Instructor: Jiang Suqi

Students: Jiao Zhiheng, Liu Tian

● **设计说明**

概念设计来源于对展览空间中"光"的认知。以一个单纯的"光"的概念作为空间生成的依据，组织展览的流线，也回应季节中光的变化和树木景观。在保持原有氛围和格局的基础上，尽量将人的活动引入地下。这是一种有节制的改造策略。概念清晰明确，转译成空间的语言比较准确，但结构略显繁杂。

● **Design notes**

The concept of this design comes from the cognition of "light" in exhibition space. Taking a simple concept of "light" as the basis of space generation, the streamline of exhibition organization also responds to the changes of light and tree landscape in seasons. On the basis of maintaining the original atmosphere and pattern, people's activities should be introduced underground as far as possible. This is a moderate transformation strategy. The concept is quite clear, and is more accurate when translated into space language, but the structure is slightly complicated.

● **概念生成**

● **Concept generation**

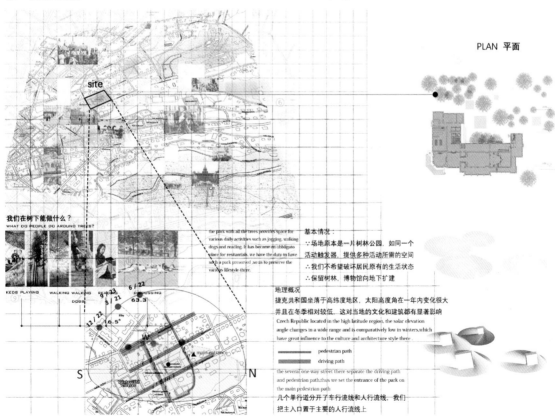

PLAN 平面

● 具体转化过程　Specific transformation process

Spinning machine

Feeding—Feeder	Comb—Combing roller	Entering—Rotor	Twisting—Lead roller	Export—Cone

Complex streamlines	**Streamline at the entrance**	**Streamline at the workspace**	**Roof shapes of workspace**	**Storage function**
• Unprocessed *cotton* into the spinning machine. • The state of mixed *flow* before entering the building.	• *Combing rollers* comb cotton into a single fiber. • *Entrance* separates crowd into flow.	• *Rotor* rotates the fiber at a high speed, and condenses into a fiber bundle, exporting through the eyelet.	• *Lead roller* rotates fibers and twists into lines.	• Cord arounds the *cone* for weaving. • Collection Storage for exhibition.

Loom

Measuring—Flat sheet	Opening—Shed	Weft insertion—Nozzle	Fabric run—Reed	Beam warping—Folder

Workspace and exhibition function	**Exhibition layout**	**Transportation space**	**Building structure**	**Envelope structure & Support structure**
• *Flat sheet* measures the length of the yarn. • *Flat sheet* is the middle step between the spinning machine and the loom. So the arc echoes the two parts in the junction of museum.	• *Yarn* through the opening of *warp*, just like people see the exhibits actively, and the exhibits are seen passively.	• Stairs connect the space.	• V-shaped *structure* system connects the entire hall as a whole.	• Lines are woven into cloth, rolling on the Heald frames. • Perforated plate skin & two lift shafts acting as bearing structures.

● 剖面图　Section

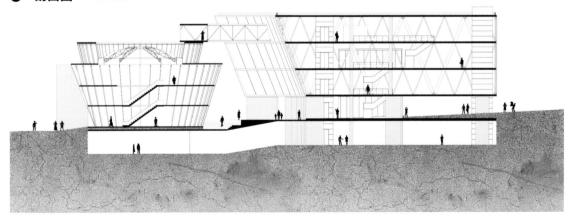

设计题目： 捷克城堡博物馆扩建设计
指导老师： 陈翠
学　　生： 李艺书、陆雨婷

Design topic: Expansion Design of Czech Castle Museum

Instructor: Chen Hui

Students: Li Yishu, Lu Yuting

18

● 效果图　Design sketch

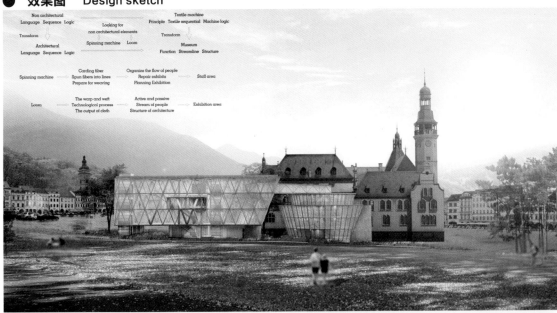

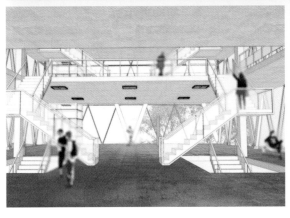

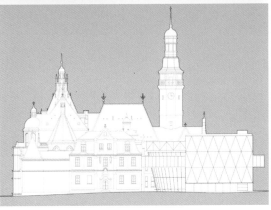

课题六：汉口胜利仓库改扩建设计
Topic 6: Reconstruction and Expansion Design of Shengli Warehouse in Hankou

● **模型效果图** Model rendering

● **光影场景** Nightscape presentation

19

设计题目： 地域建筑 —— 长沙造纸厂工业改造与更新设计
获奖名称： 2016 年中国建筑院校境外交流作业展二等奖
指导老师： 罗荩、陈翚
学　　生： 何磊

Design topic: Regional Architecture—Industrial Transformation and Renewal Design of Changsha Paper Mill

Award: Second prize of 2017 overseas exchange operation exhibition of Chinese architecture colleges

Instructor: Luo Jin, Chen Hui

Student: He Lei

● **设计概念**

灵感来源于工厂建筑剩余的封闭式和部分封闭的庭院、
露台、阳台空间，思考了中国南方地区建筑的庭院和"露
台"（天井）关系，并结合中国南方地区建筑的地域特点，
进行工业改造和更新设计。

Design notes

Inspired by the remaining enclosed and partially enclosed courtyards, terraces and balconies of workshop buildings, this paper considers the courtyards and terraces of buildings in southern China, and carries out industrial transformation and renewal design in combination with the regional characteristics of buildings in southern China.

● **首层平面　Ground floor plan**

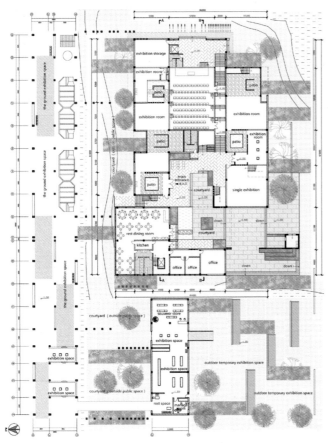

● **剖面图　Profile**

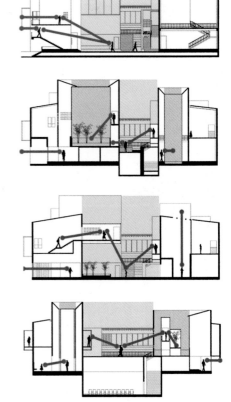

● **夜景效果图** Nightscape presentation

● **光影场景** Light and shadow scene

20

设计题目： 2018 汉口胜利仓库改造设计
指导老师： 张蔚
学　　生： 李敬萱、任意

Design topic: Reconstruction Design of Shengli Warehouse in Hankou in 2018

Instructor: Zhang Wei

Students: Li Jingxuan, Ren Yi

● **设计说明**

场地内部缺乏活力，通过架设天桥的方式将周围人流较大的地点连通，并将人群引入场地，汇聚活力。利用新老结构的结合营造码头的场所与氛围感。

Design notes

There is a lack of vitality inside the site. The places with large flow of people around are connected by erecting overpasses, and the crowd is introduced into the site to gather vitality. The combination of new and old structures is used to create a sense of place and atmosphere of the wharf.

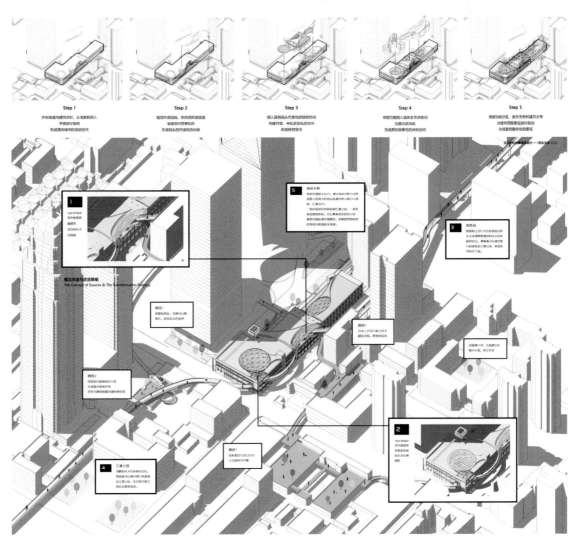

● 场景透视　Scene perspective

● 空间注记　Space notes

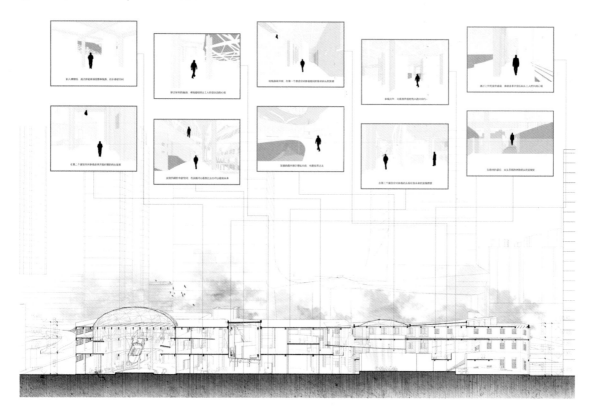

● 一层平面图　First floor plan

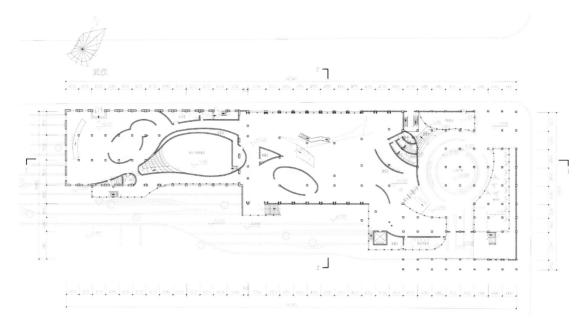

● 结构分析　Structural analysis

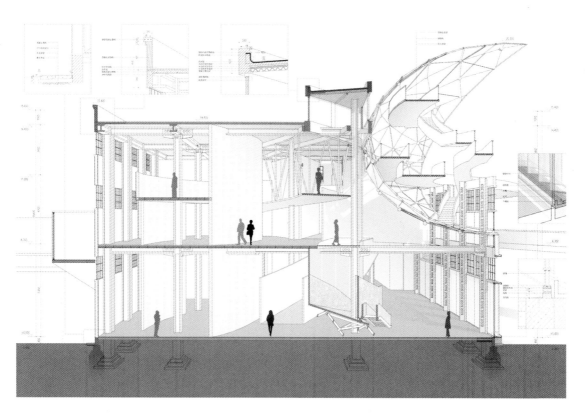

四年级

Fourth grade

四年级建筑设计课程介绍
Course Introduction of Architectural Ddesign Course for Grade Four

课程内容：建筑设计 V、建筑设计 VI

Course content: Architectural Design V, Architectural Design VI

教师团队
Teacher team

王小凡
Wang Xiaofan

徐峰
Xu Feng

卢健松
Lu Jiansong

袁朝晖
Yuan Chaohui

邓广
Deng Guang

陈晓明
Chen Xiaoming

齐靖
Qi Jing

宋明星
Song Mingxing

严湘琦
Yan Xiangqi

刘尔希
Liu Erxi

课程介绍
Course introduction

本课程是建筑学专业本科生的主干专业课程，包括建筑设计 V 和建筑设计 VI。建筑设计 V 课程通过讲授高层建筑设计基本理论及知识，培养学生的实践能力和正确的设计方法。该课程要求掌握高层建筑及写字楼设计创作的一般规律与方法，对高层建筑与城市周围环境的关系有一定的认识。此外，还要了解高层建筑基本原理及防火规范，了解结构体系及选型，对高层建筑的造型和结构有一定的整体把控，充分了解高层建筑在城市中发挥的作用。

建筑设计 VI 是建筑学本科四年级第二门专业设计课程。该课程通过传授大跨度建筑设计基本理论及知识，培养学生的实践能力和正确的设计方法，从而从流线、结构选型、大空间平面选型及建筑物理技术条件来解决该类建筑的空间构成、技术构成、结构选型和消防等问题，学生在掌握设计基本原理之外，还要通过模型制作及计算机辅助设计等手段达到掌握大跨度建筑设计的技巧和表现方法，加强学生制作比较复杂的工作模型的能力和对较复杂建筑类型的理解以及大跨度建筑的建筑设计原理的学习。通过对城市实地调研，加强学生对城市需求及社会化复杂性的理解。

This course is the main professional course for undergraduates majoring in architecture,including architectural design V and architectural design VI. Architectural design V aims to cultivate students' practical ability and help them learn design methods by teaching the basic theory and knowledge of high-rise building design.This course requires students to master the general laws and methods of high-rise building and office building design,and have a certain understanding of the relationship between high-rise buildings and the surrounding environment of the city.Besides,students should understand the basic principles and fire prevention codes of high-rise buildings,the structural system and type selection,and have a certain overall control over the modeling and structure of high-rise buildings,and then fully understand the role of high-rise buildings in the city.

Architectural design VI is the second professional design course for undergraduates in the senior year of architecture.By imparting the basic theory and knowledge of wide-spanning architectural design,this course cultivates students' practical ability and help them learn design methods,and students are projected to solve the problems, including space and technical composition,structure selection and fire protection of such buildings from the streamline,structure selection,free plan selection and physical and technical conditions of architecture.In addition to mastering the basic principles of design，students should also master the skills and methods of expressing wide-spanning architecture design by means of model-making and computer-aided design,and strengthen their ability to make more complex working models and understand more complex architecture types,and learn the architectural design principles of wide-spanning architecture.This course intend to improve students' understanding of urban needs and the complexity of socialization through on-the-spot investigation of cities.

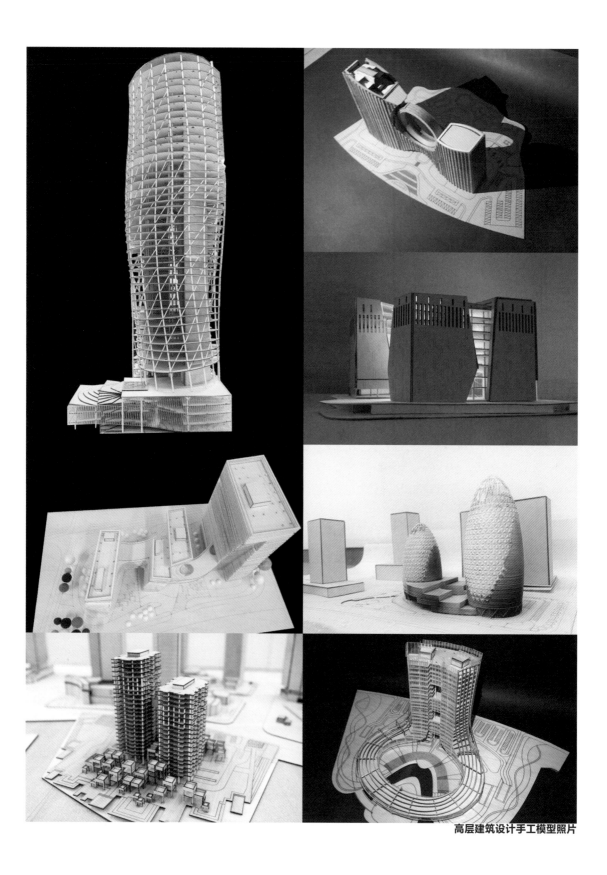

高层建筑设计手工模型照片

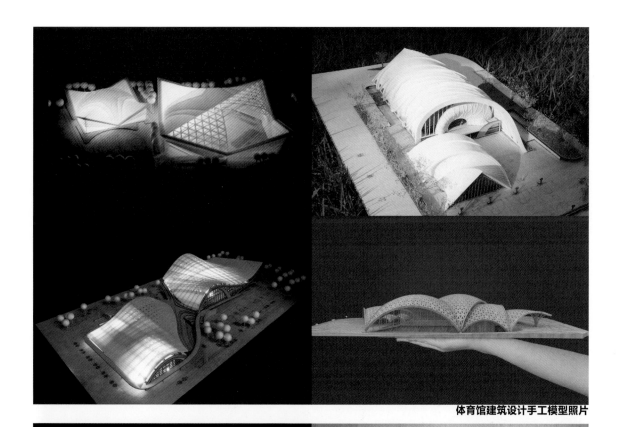

体育馆建筑设计手工模型照片

剧院建筑设计手工模型照片

高层建筑设计
Topic 1: High-Rise Building Design

一、主题

建筑设计 V 传授高层建筑设计基本理论及知识，培养学生的实践能力和正确的设计方法，培养学生对高层建筑与城市周围环境的关系的敏锐认知，从而从功能、环境、规范及技术条件来解决高层建筑的空间构成、环境创造、结构选型和消防等问题。学生在掌握设计基本原理之外，还要通过绘图、模型制作及计算机模拟辅助设计等手段达到熟练掌握高层建筑设计的技巧和表现方法。

二、教学内容及要求

1. 设计内容

掌握高层建筑及写字楼设计创作的一般规律与方法；对高层建筑与城市周围环境的关系有一定的认识；了解高层建筑基本原理及防火规范；了解结构体系及选型。

通过实地调研、课程设计作业，加强学生对技术、经济、法规、建筑文化的理解以及对高层建筑设计原理的学习；同时通过增加快题设计量，加强学生动手能力。

2. 设计要求

（1）掌握高层建筑的定义、分类和发展简史；了解高层建筑相关背景和热点问题；了解高层建筑基于生态的设计侧重。

（2）了解高层建筑生态设计的基本方法。

（3）掌握生态软件模拟的方法和工具。

（4）掌握高层建筑总平面设计的基本方法、标准层设计的基本方法、高层建筑设备层、结构转换层设计的基本方法、高层建筑防火设计规范。

（5）掌握建筑设计的表达与基本表现方法、掌握手工模型的制作方法。

大跨度建筑设计
Topic 2: Large Span Architectural Design

一、主题

建筑设计 Ⅵ 是建筑学专业本科生的主干专业课程。该课程通过传授大跨度建筑设计基本理论及知识，培养学生的实践能力和正确的设计方法，从而从流线、结构选型、大空间平面选型及建筑物理技术条件来解决该类建筑的空间构成、技术构成、结构选型和消防等问题。学生在掌握设计基本原理之外，还要通过模型制作及计算机辅助设计等手段，掌握大跨度建筑设计的技巧和表现方法。

二、设计内容及要求

1. 设计内容

掌握中小型体育馆的功能构成及流线组织；熟悉大空间中视线升起分析及了解声学分析方法；掌握大空间建筑的安全疏散及消防设计，了解大空间组合与结构选型规律；了解各类体育项目场馆的不同需求及尺度。掌握中小型剧院的功能构成及流线组织；了解各类剧场、音乐厅的不同需求及尺度；了解大空间组合与结构选型规律；尝试结合校园周边环境、现有地形进行设计。

2. 设计要求

（1）体育馆建筑外部形象是校园整体形态的一部分，也是城市界面的一部分，建筑设计必须考虑与校园环境、城市道路、周边环境之间的关系。本建筑方案应满足体育馆基本要求，同时可考虑演出、展览及非教学时间段广大师生及周边城市居民的多元使用要求。

（2）剧院设计主要功能包含开展影剧院观演及排练等。综合考虑观众、演员、贵宾、布景制作、工作人员等多种流线。设计需考虑基本的声学设计和视线升起要求。要求结合周边环境因素，选择相应的大空间结构形式开展设计，精心布置停车场、绿化、雕塑等。应侧重于从建筑结构创新、建筑文化传达、建筑与城市的关系、数字化设计等主题方向开展设计。

（3）图纸表达规范：图纸能充分表达出作品创作意图，并且包含设计概念描述和设计说明。

（4）要求制作精细手工模型。

课题一：高层建筑设计
Topic 1: High-Rise Building Design

设计题目： 野蛮生长 —— 城市商业综合体设计
指导老师： 徐峰、袁朝晖
学　　生： 杨明达、涂欢乐

Design topic: Wild Growth—Urban Commercial Complex Design

Instructors: Xu Feng, Yuan Zhaohui

Students: Yang Mingda, Tu Huanle

● **设计说明**

这个城市综合体的设计没有概念先行，而是选择基于对场地充分且理性地解读并进行不断深入的设计推演，最终生成一栋在南门口老城区中呈现出野蛮生长状态的建筑。该综合体集日常百货、商业零售、人才公寓与艺术展览于一体，同时为年轻游客、青年人、老住区居民服务，致力于增强空间与城市的联结，并提升用户体验。

Design notes

The design of this urban complex does not act on concept but chooses to conduct in-depth design and inferences based on sufficient and rational interpretation of the site,and finally a building showing a state of wilde growth in the old town of South Gate generates.The complex combines department stores,commercial retail,talent apartments and art exhibitions while serving young visitors, young people and people living in old neighborhoods.It is a great way to connect the space with the city and improve the user experience.

■风环境分析 Wind Environment analysis

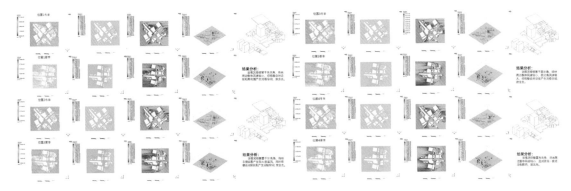

■多首层分析 Multi-layer analysis

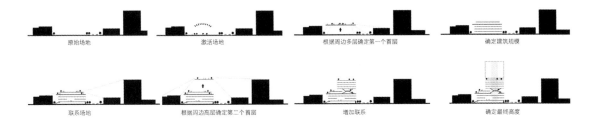

■场地分析 Site analysis

功能区块

图底关系

内部交通

周边交通

景观通廊

人群分布

■区位分析 District analysis

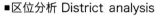

■核心空间分析 Core space analysis

我们将小型的商业功能植入人群活动最复杂的位置，使得整个空间独立于为三种类型人员服务的专属空间，形成第四种类型的空间组团，成为整个建筑的"心脏"——交通枢纽，向之前所提的三种不同类型的空间输送"血液"——人流。

在手法上，我们使用拆解九宫格的方式，进行空间的分割与组合，试图模拟立体的街道空间，同时暴露结构，以还原老长沙街巷中野蛮生长的氛围。

灰空间体块   中庭空间体块

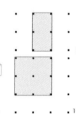
8F   10F

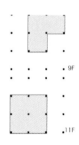
9F   11F

■体块生成分析 Volime generation analysis

规划空间与流线   多首层商业模式   整理交通，为三种人群提供便捷的出行方式

增加空中酒吧吸引人流   向商业街展开   引导性的交通

整合现有资源，以根植城市文脉的城市更新盘活老旧街区，沟通商业地段，实现城市全面升级，是设计前期思考的核心。城市综合体不仅在物层面上衔接老城区与商图，而且是能实现新旧城区的深层次对话，并以完善的城市功能为人才交流、居民生活提供宜人的栖居地。

Integration of existing resources,urban renewal rooted in urban history, revitalization of old blocks, communication of commercial areas, and the realization of comprehensive urban upgrading are the core of the design in the early stage.Urban complex not only connects the old city and the business districts on the surface, but can also realize the deep interaction between the new and old city blocks,and provides convenient conditions for the exchange of talents and the life of residents with perfect urban functions.

■人群分析 Crowd analysis

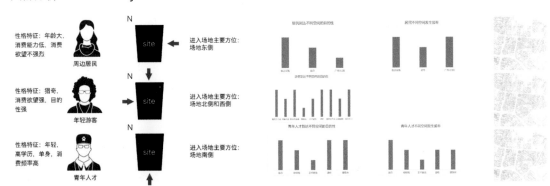

■行为分析 Behavior analysis

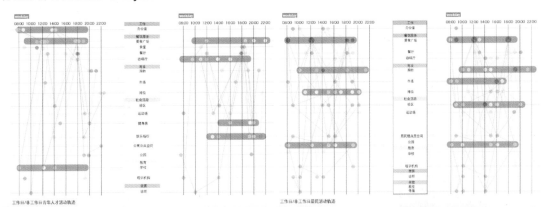

设计题目:"湘"笋 —— 绿色高层建筑设计
指导老师:徐峰
学　　生:胡婉婷、夏心玉

2

Design topic: "Xiang" Bamboo Shoots—Design of Green High-rise Buildings

Instructors: Xu Feng

Students: Hu Wanting, Xia Xinyu

● 设计说明

本双塔楼坐落于湘江西畔,该设计以梭形圆锥体为基本型,形态独特挺拔,流线型外观犹如雨后春笋,寓意着长沙未来金融中心的生机勃发,为长沙滨江商务带增添独特的风景线。塔楼平面以圆形为基本型,最大化利用平面空间,室内可以沿玻璃幕墙尽享360°自然采光,室外则一览湘江及滨江景观带多重景色,提升空间环境的舒适度。立面遮阳表皮由可开启的三棱锥单元牵动PTEF膜组成,可电动控制,最大化完成各个角度遮阳及热辐射控制。不同立面的表皮根据太阳角度在不同时间段开启,夜晚构件收缩,人们的视线可达到360°全开放,完成遮阳与赏景的完美切换。裙房由圆形平面层层叠加而上,形态轻盈而平稳地托起两座塔楼,犹如一叶扁舟泊在湘水西畔;南面架空的底层与中庭贯通,让人们身处裙房内也能感受到凉爽的夏风。

Design notes

The twin towers are located on the west bank of the Xiangjiang River.Its design takes the shuttle cone as the basic shape,with unique and straight shape and streamlined appearance, which implies the vitality of Changsha's future financial center and adds a unique landscape to Changsha's riverside business belt. The plane of the tower is basically circular,making maximum use of the plane space.People indoor can enjoy 360 degrees of natural lighting along the glass curtain wall,while people outdoor can take in multiple views of the Xiangjiang River and the riverside,so the design improves the comfort of the environment.Shading facade skin is composed of openable triangular pyramid unit affecting PTEF film,which can be electrically controlled to maximize the sunshade and heat radiation control at all angles.The skins of different facades are opened at different time periods according to the sun angle and people's line of sight can be fully opened at 360 degrees to complete the perfect switching between sunshade and landscape viewing when the components shrink at night.The podium is stacked layer by layer from the circular plane and holds up the two towers lightly like a boat moored on the west bank of Xiangjiang river.The overhead ground floor in the south is connected with the atrium,so that people can feel the cool summer wind in the podium.

● 区位分析　Location analysis

● 交通功能分析　Traffic function analysis

● 风环境分析　Wind environment analysis

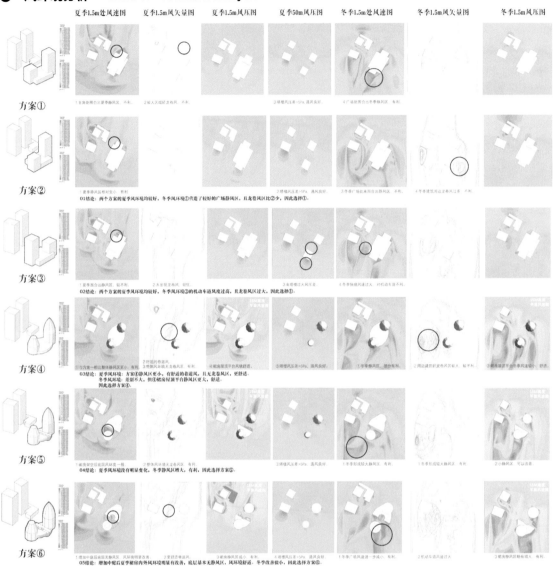

夏季1.5m处风速图　夏季1.5m风矢量图　夏季1.5m风压图　夏季50m风压图　冬季1.5m处风速图　冬季1.5m风矢量图　冬季1.5m风压图

方案①　方案②　方案③　方案④　方案⑤　方案⑥

运用 CFX 风环境模拟结果，先对比分析确定两座塔楼的高低关系与平面对位关系，再确定塔楼的基本形态，最后确定裙房的高低关系以及是否设置中庭。通过对 6 个方案夏季、冬季风环境的模拟比较，最终选择方案 6 作为本课程设计的推进方案。

We compare and analyze the height relationship and plane alignment relationship of the two towers by the CFX wind environment simulation results,and then determine the basic shape of the tower,and finally the height relationship of the podium and whether to set the atrium.Simulating and comparing the wind environment of six schemes in summer and winter,we finally selected the sixth scheme as the promotion one of the course design.

● **总平面图**　General plan

● **爆炸图**　Exploded diagram

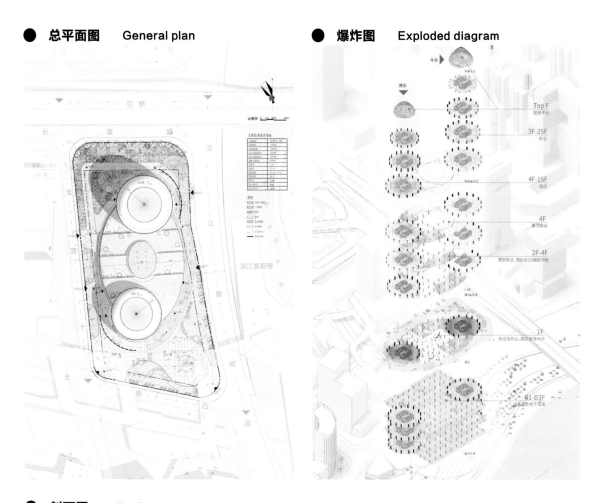

● **剖面图**　Profile

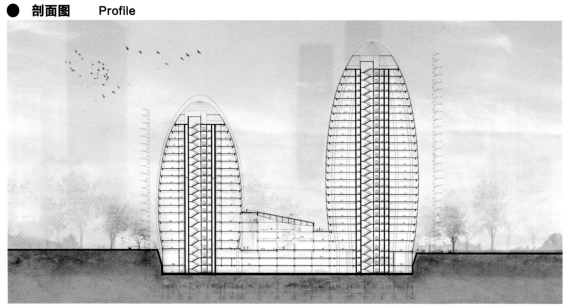

设计题目： 湖滨有个圆 —— 绿色高层建筑设计
指导老师： 卢健松
学　　生： 李敬萱、彭丹、冉富雅

3

Design topic: There is a Circle on the Lakeside—Design of Green High-rise Buildings

Instructors: Lu Jiansong

Students: Li Jingxuan, Peng Dan, Ran Fuya

● 设计说明

本设计脱离了高层塔裙分开设计的普遍手法，运用圆的流动性和外表皮的扭转，使塔楼和裙房紧密地结合在一起，弱化裙房边界，让整栋建筑更加整体，形成了简洁的形态和丰富的空间。中间部分的场地采用大台阶逐步将人引到湖边，同时改变了原场地规整的布局，将湖水引入建筑内部，设计了景观道，使得室内外都有着良好的视线，增加了人们的亲水行为。通过绿建软件模拟辅助设计建筑的朝向、遮阳等部分，使建筑具有一定的绿色可持续性。

Design notes

Breaking away from the common method of separating design of high-rise towers and skirts, this design uses the fluidity of circle and the torsion of the outer skin to closely combine the tower with the podium, which weaken the boundary of the podium and make the whole building more integral, so that forms a concise form and enough space. The site in the middle part adopts large steps to gradually lead people to the lakes. At the same time, the regular layout of the original site is changed and the lake water is introduced into the interior of the building, and the landscape roads here provide a good line of sight indoors and outdoors and people's hydrophilic behavior is increased. The building has green sustainability through the green building software to simulate the orientation and shading of the auxiliary design building.

● 区域分析　　Regional analysis

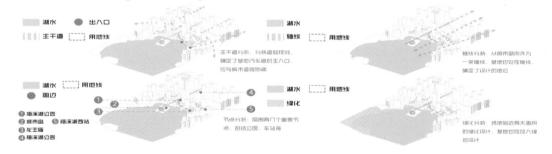

● 方案演绎　　Scheme deduction

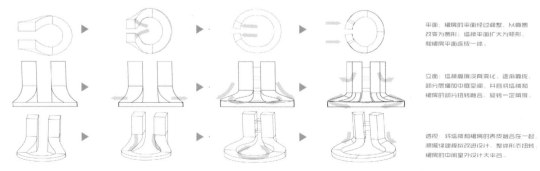

● 形态生成　Morphogenesis

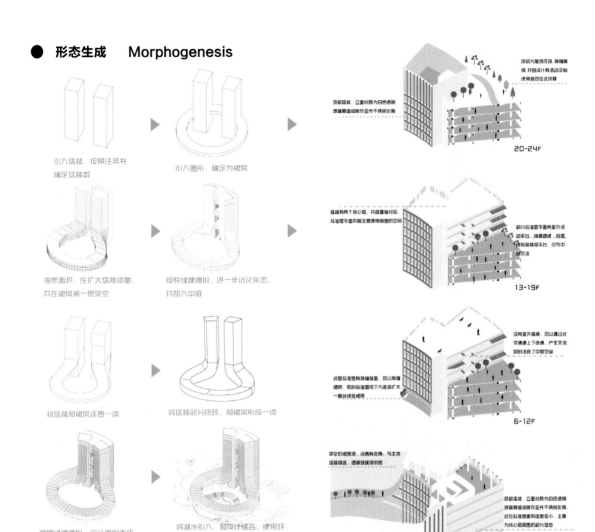

引入塔楼，按照任务书
确定塔楼数

引入圆形，确定为裙房

考虑面积，在扩大塔楼体量，
并在裙房第一层架空

按照绿建模拟，进一步优化形态，
并加入中庭

将塔楼和裙房连贯一体

将塔楼部分扭转，和裙房形成一体

根据绿建模拟，设计遮阳表皮

将湖水引入，和设计结合，使用环
桥，作为消防扑救

顶部为屋顶花园，种植植
被，并且设计有活动设施，
使得着归可在此流憩

顶部塔楼，立面制质为白色透明
玻璃幕墙和哑光不锈钢龙骨

20-24F

塔楼有两个核心筒，并且置楼对称，
标准层平面布置主要使用周围的空间

部分标准层平面有室外活
动平台，种植植被，后层，
标准楼梯相平台，可与中
庭交流

13-19F

没有室外楼梯，可以通过此
交通连上下连通，户主交流，
同时丰富了中庭空间

此层标准层有种植植面，可以种植
植被，同时标准层向下迅速扩大
一直延伸呈裙房

6-12F

架空的裙房底，设置有龙骨，与主体
塔楼相连，透明玻璃顶材质

底部塔楼，立面制质为白色透明
玻璃幕墙和哑光不锈钢龙骨，
此处标准层面积逐渐变小，主要
为核心筒周围的部分墙墙

1-5F

● 人视图　Personal view

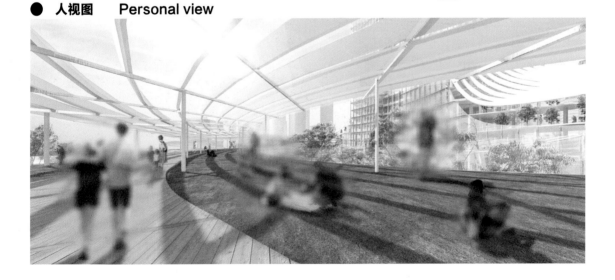

158

● 景观及节点场景　Landscape and node scene

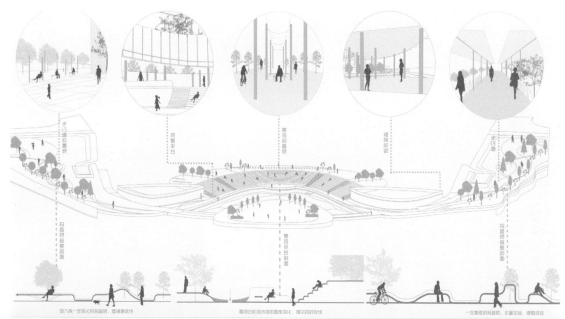

● 手工模型照片　Manual model photos

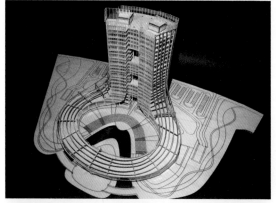

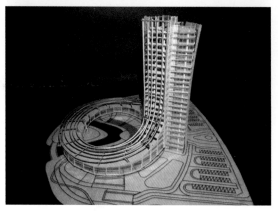

160

设计题目： 浮塔 —— 建筑设计 V 高层建筑设计

指导老师： 严湘琦、王小凡

学　　生： 李杼欣、高原

4

Design title: Suspension Tower—Architectural Design V High-rise Building Design

Instructors: Yan Xiangqi, Wang Xiaofan

Students: Li Zhuxin, Gao Yuan

● 场地分析

长沙南湖金融片区西临湘江河畔，该地段公共交通便利，以平安金融、保利国际和汇景发展为主的 200m 以上写字楼成为沿江城市天际线。基地地处滨江地段，作为政府规划的金融前台片区，其超高层需承担城市名片的价值属性。Suspension Tower 不是一栋纯功能性的高层建筑，从差异化优质办公空间出发，借助幕墙媒介交互技术，为新兴独立金融公司打造其产业形象与影响效益，同时实现沿江城市风貌。11 个多层办公单体悬挂，屋顶花园＋通高桁架层＋差异化标准层，增加了不同业主使用空间的多样性，实现超高"多层"。悬吊结构带来的单体无柱空间让办公层视野最大化。

Site analysis

Adjacent to the Xiangjiang River in the west,Changsha Nanhu financial area,with convenient transportation,becomes the city skyline along the river with the office buildings of more than 200m focusing on Ping' An Finance,Poly International and Huijing Development. Located in the riverside area,the base is taken as an important financial area planned by the government,and its super high-rise needs to show the value of the city.Starting from a differentiated high quality office space, Suspension Tower is not only a purely functional high-rise building but also create its industrial image and impact benefits for emerging independent financial companies while realizing the city style along the river with the help of curtain wall media interaction technology.With eleven suspended multi-storey office units,the roof garden,high-level truss layer,differentiated standard layer increase the diversity of the space used by different owners and realizes ultra-high "multi-layer".Suspended structure in the single column-free space maximizes the vision of the office.

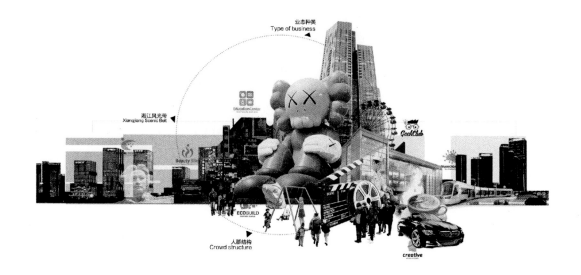

● 功能分析　Function analysis

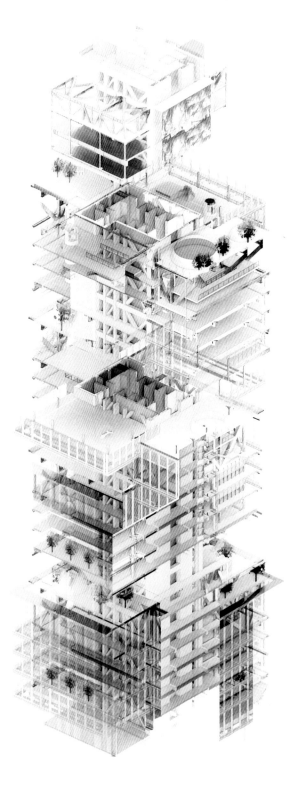

单体屋顶花园 GARDEN

空中连廊 CORIDOR

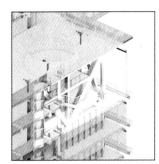

通高桁架层 RECREATION

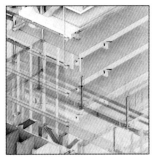

差异化办公层 WORK

设计题目：活力聚合链 —— 长沙南湖金融区高层设计
指导老师：卢健松、陈晓明
学　　生：陈思奇、武文忻

Design topic: Dynamic Aggregation Chain—High-rise Design of Changsha Nanhu Financial District

Instructors: Lu Jiansong, Chen Xiaoming

Students: Chen Siqi，Wu Wenxin

5

● 设计说明

高层是城市发展的一个象征，它不但能够节省土地，而且能够减少政府对公共设施的开发周期和投资。高层建筑有加快城市建设的优点，但同时也带来了种种问题，如安全问题、公共卫生问题、社会问题等。人们在高层里似乎也渐行渐远，脱离了自然与生态。

Design notes

High-rise building is a symbol of urban development, which can not only save lands, but also reduce the government's development cycle and investment in public facilities.High-rise buildings have the advantage of speeding up city construction, but they also bring various problems,such as safety problems,public health problems, social problems and so on.People seem to be drifting away from nature and ecology in the high buildings.

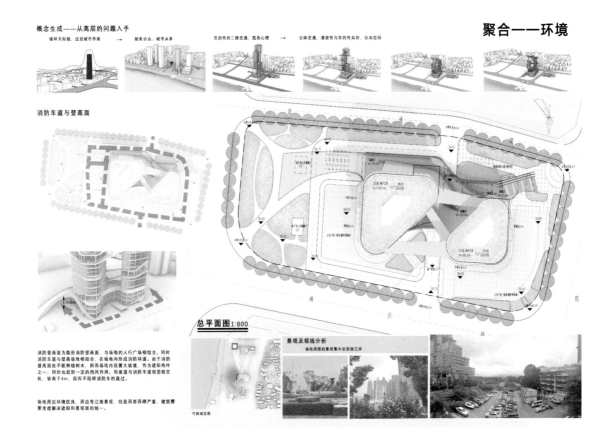

本课程设计的概念为"活力聚合链"，意在用空间廊道关系将高层的人串联起来，使得人们在工作之余，也可以享受在路上漫步的休闲时光。方案将整个坡道分为三段在特定区域联系的交通体系，其他坡道空间结合遮阳与建筑要求，布置相应的功能。

同时，方案意在以一种生态的、自然的方式构建高层双塔体系。因而方案使用了一些绿色建筑的手法，对建筑的风环境、光环境进行分析，并据此生成遮阳构件，同时也为坡道的植被种植选择提供依据。

高层越来越高，破坏天际线，压迫城市界面，破坏整体空间关系。在城市关系中，高层应服务公众，使得地坪广场成为城市公共场所的一部分。

The concept of this course design is "dynamic aggregation chain",which aims to connect people in the high-rise buildings with the spatial corridor so that people can enjoy the leisure time of walking on the road after work.The plan divides the entire ramp into three sections of traffic system in specific areas,while other ramp spaces are arranged with corresponding functions in combination with shading and architectural requirements.

The concept of this course design is "dynamic aggregation chain",which aims to connect people in the high-rise buildings with the spatial corridor so that people can enjoy the leisure time of walking on the road after work.The plan divides the entire ramp into three sections of traffic system in specific areas,while other ramp spaces are arranged with corresponding functions in combination with shading and architectural requirements.

At the same time, the plan aims to build a high-rise twin towers in an ecological and natural way.The scheme uses some green building techniques to analyze the wind and light environment of the building.Based on this, shading construction is generated, and it also provides a basis for the selection of vegetation planting on the ramp.High-rise buildings destroy the skyline,oppress the urban interface and destroy the overall spatial relationship.In the urban relationship,the high-rise buildings should serve the public,and make the terrace square a part of the urban public places.

● 风环境分析

从夏季、过渡季风主导风向为东南风的情况下距地面高度1.5m处人行高度处风速矢量分布圈可以看出：项目建筑规划布局合理，建筑间距控制适当，不存在涡流，场地内人活动区无静风区，满足规范要求。由夏季、过渡季风向为东南风的情况下距地面1.5m处人行高度的风速云图可知，周边人行区域距地1.5m高度处最大风速分别为4.86m/s，可以看出：人行高度处的风速基本处于合理的范围之内，人行高度处的风速基本处于0.76m/s~3.42m/s。部分范围内风速较大，但不会影响人们的室外活动。最后，从不同区域1.5m水平面风压云图可以看出：迎风侧高层建筑前后表面静压差均在5Pa以上，有利于夏季室内自然通风。参评建筑开启外窗内外表面风压差均大于5Pa，符合规范标准。

Wind environment analysis

From the wind speed vector distribution circle at the pedestrian height 1.5m from the ground height in summer, when the dominant wind direction of transition monsoon is southeast wind, it can be seen that the construction planning and layout of the project is reasonable, the spacing between buildings is properly controlled, there is no eddy current, and there is no calm wind area in the human activity area of the site, which meets the requirements of the code. According to the wind speed cloud map of the pedestrian height 1.5m above the ground in summer and the south-easterly direction of the transitory monsoon, the maximum wind speed of the surrounding pedestrian area at the height 1.5m above the ground is 4.86m/s respectively. It can be seen that the wind speed at pedestrian height is basically within a reasonable range, and the wind speed at pedestrian height is basically within the range of 0.76 mL /s ~3.42m /s. Wind speed is larger in some areas, but it will not affect people's outdoor activities. Finally, the wind pressure cloud map of 1.5m horizontal plane from different districts. It can be seen that the static pressure difference between the front and rear surfaces of windward high-rise buildings is more than 5Pa, which is conducive to indoor natural ventilation in summer. The wind pressure difference between the exterior and exterior surfaces of the open Windows of the participating buildings is greater than 5Pa, which is in line with the standard.

夏季风环境

  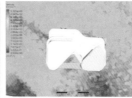 

距地面1.5m风速矢量图　　　　　　　　　　风速云图　　　　　　　　　　各区域距地面1.5m风速放大系数图

● 手工模型照片

Manual model photos

 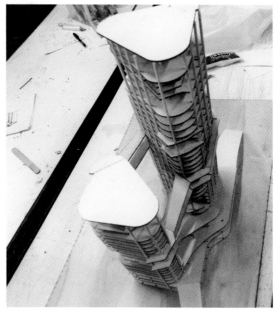

6

设计题目：引渡缝隙 —— 绿色高层建筑设计
指导老师：刘尔希
学　　生：李思佳 、王媛

Design topic: Extradition Gap—Design of Green High-rise Buildings

Instructor: Liu Erxi

Students: Li Sijia, Wang Yuan

● **设计说明**

基于前期对基地区位的梳理，如何利用它的景观优势，并将景色渗透到建筑之中成为我们的设计重点。结合人流的趋向性，我们将基地东西向贯通，以更加主动的姿态刺激人流流动，在与东侧绿地自然环境的交互中实现"开放街区"。形体上寻求简洁与力量感，以错落倾斜的墙体暗示通道的存在，借助光影偏移将形体的内斜倾向与"取景格栅"相呼应，制造移步换影的特殊空间感，营造趣味性。表皮的意象来源于"岩岩缝隙"，其中"明"角由亮面玻璃构成，"暗"角由外立面掀起的百叶渐变而成，在简洁几何形体上呈现出明暗交错的节奏与韵律。错落的几何形态和倾斜相交的柱网支撑，给人带来特殊的动态感。钢架的运用则扩大了室内的活动空间，便于灵活自由地适应多种功能活动的布置。

● **Design notes**

Based on the combination of the location of the base in the early stage,how to make use of its landscape advantages and infiltrate the scenery into the architecture has become our design focus. Combined with the trend of people flow,we connect the base from east to west to stimulate people flow with a more active attitude,and realize "open block" in the interaction with the natural environment of the green space on the east side.Seeking simplicity and power in the form,it hints the existence of channels with scattered and inclined walls,and echos its inclination of the form with the "viewfinder grid" with the help of light and shadow offset,so as to create a special sense of space for step-by-step shadow exchange and create interest.The image of the surface comes from "the cracks in rock",in which the "bright" corner is composed of bright glass,and the "dark" corner is gradually formed by the louvers raised by the facade,showing the rhythm of the crisscross of light glasses and dark louvers on the simple geometric form.The scattered geometric shape and the obliquely intersecting column grid support bring a special dynamic sense to people. The use of steel frame expands the indoor activity space so that people can flexibly and freely adapt to the layout of multi-functional activities.

● **场地分析　Site analysis**

● 环境模拟分析　Environmental simulation and analysis

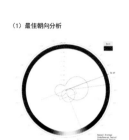

（1）最佳朝向分析

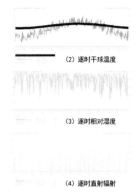

（2）逐时干球温度

（3）逐时相对湿度

（4）逐时直射辐射

（5）风环境分析

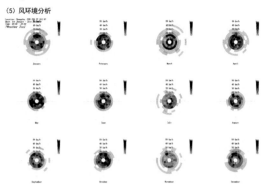

● 形体生成　Shape generation

利用场地地位置，由商业区到江滨带过渡
Use the site as the transition
commercial area to landscape

顺应轴线，抬升片状体块形成引导性通道
Following the axis, a leading
slit channel forming

南向形成退台，进行自然采光的引入
Adjust the height of the block,
introduce natural lighting

南侧形体扭转形成喇叭口，鼓励活动
Twisted to form a bell mouth toward the
green space, encourage people to move

北侧形体收合，景观面集中于东侧绿地
Concentrated the landscape in the eastern
green space and waterscape

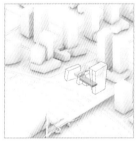

形体连接打破独立性，统一建筑体
Insert corridor connecting form, break
the independence and unify building

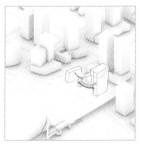

空中连廊调整为弧形，增加形体动态
Adjust the corridor in the air to expand
the arc, increase the body dynamics

根据场地规范与面积调节，形成最终形态
Adjust according to site specification
and building area to final form

● 手工模型　Manual model

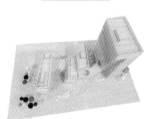

● 绿色廊道　Green corridor

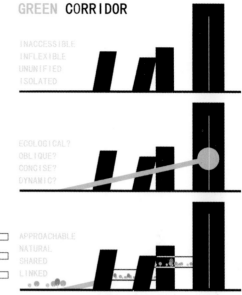

GREEN CORRIDOR

INACCESSIBLE
INFLEXIBLE
UNUNIFIED
ISOLATED

ECOLOGICAL?
OBLIQUE?
CONCISE?
DYNAMIC?

APPROACHABLE
NATURAL
SHARED
LINKED

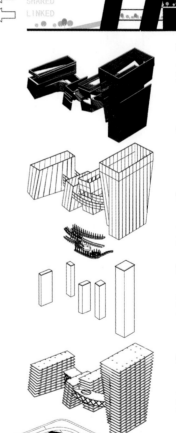

BUILDING SKIN

——一体化的外加百叶根据内部功能与立面朝向，以城隙的尺寸调节室内光照与温度

GLASS SCREEN

——以生态玻璃分隔覆盖，利用周围良好的景观优势，提供更加广阔的视野，营造轻透感

THE CORRIDOR

——内部连廊设置生态廊道，弧形及旋转拾升的坡道为观测的办公空间插入了更加轻松自然的环境

TRAFFIC SPACE

——内部交通主要由核心简与集中式楼梯间承载，提高利用效率，降低交通引起的空间浪费

CONSTRUCTION

——倾斜扭转结构由D750的斜柱进行支撑，空中廊道由外加桁架固定，薄弱处由横向桁架梁进行加固

SITE LANDSCAPE

——场地景观设计顺应建筑形体，用不同的铺装强调了连廊下空间的活跃性，在满足足规范的基础上增加绿化，作为地面景观向东转型景观带的缓冲

7

设计题目： HOPSCA—— 绿色高层建筑设计
指导老师： 王小凡
学　　生： 蔡雨桐、黄玉洁

Design topic: HOPSCA—Green High-rise Building Design

Instructor: Wang Xiaofan

Students: Cai Yutong, Huang Yujie

● 设计说明

从摩天大楼的标准体量开始研究，设计利用一个直截了当的设计策略将典型平面的高效性和经济性最大化，同时兼顾与周边环境的呼应。由于建筑是被放在一个很小的、没有个性的、被其他高层建筑包围着的受限制的基地上，建筑的形体设计与周边环境建立了较强的联系。通过将建筑北侧的体量向南侧扭转，一个尖锐的建筑形体却给建筑的底部带来了一种全新的城市尺度，这种近似屈从的姿态重组了标准的建筑体量，并为大厦的使用者和场地附近的市民提供了一种更丰富的环境感受。项目设计综合了摩天大楼尺度的高效性以及对人体尺度舒适性的极大关注，使二者都达到了最佳。建筑底层尖锐化的处理手法弱化了建筑体量的巨大，同时提供了一种强有力的、标志性的友好建筑形象。

Design notes

Starting with the standard mass of a skyscraper, the design utilizes a straightforward design strategy to maximize the efficiency and economy of a typical plan, while taking into account the response to the surrounding environment. Because the building is placed on a small, impersonal, restricted base surrounded by other high-rise buildings, the physical design of the building establishes a strong connection with the surrounding environment. By twisting the volume of the north side of the building to the south side, a sharp building shape brings a new urban scale to the bottom of the building, which reorganizes the standard building volume and provides a richer environmental feeling for the users of the building and the citizens near the site. The project design combines the efficiency of the huge scale of the skyscraper with the great attention paid to the human scale, making both optimal. The sharpening of the building floor weakens the sheer size of the building while providing a strong, iconic, friendly architectural image.

● 场地分析与形体生成　　Site analysis and shape generation

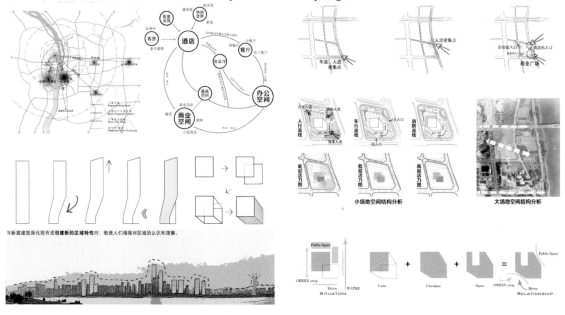

● 消防设计　Fire protection design

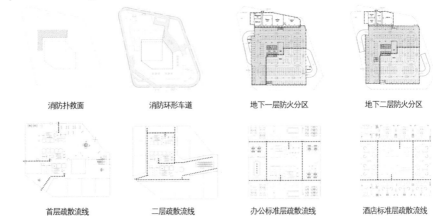

消防扑救面　　　消防环形车道　　　地下一层防火分区　　　地下二层防火分区

首层疏散流线　　　二层疏散流线　　　办公标准层疏散流线　　　酒店标准层疏散流线

● 幕墙分析　Curtain wall analysis

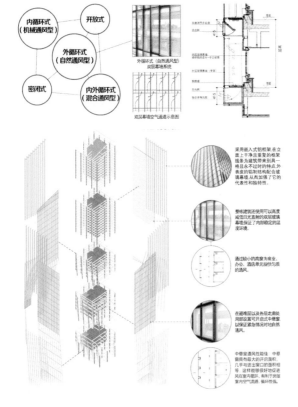

● 指标计算　Index calculation

● 剖面设计　Section design

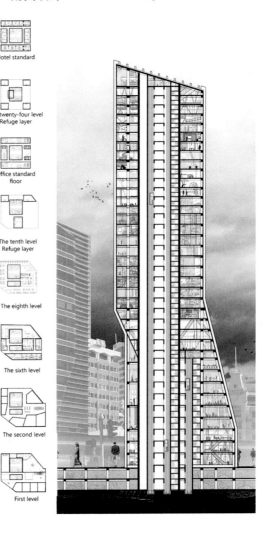

Hotel standard

The twenty-four level Refuge layer

Office standard floor

The tenth level Refuge layer

The eighth level

The sixth level

The second level

First level

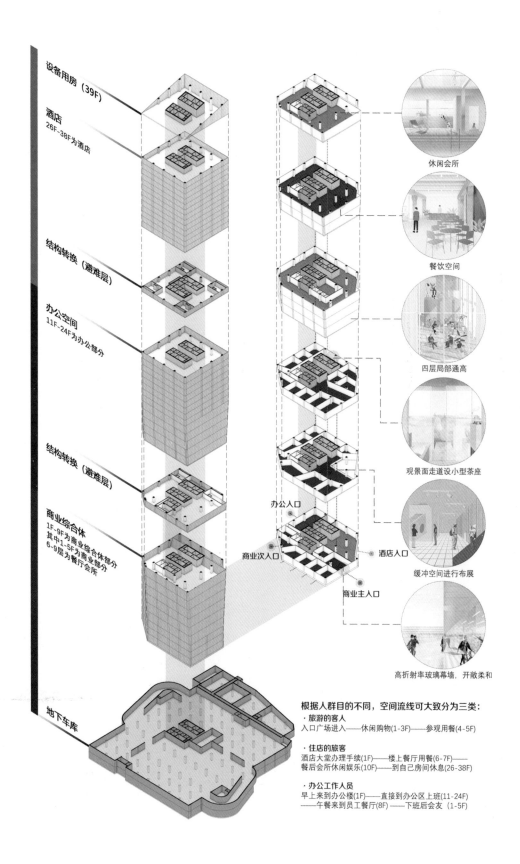

设备用房（39F）

酒店
26F-38F为酒店

结构转换（避难层）

办公空间
11F-24F为办公部分

结构转换（避难层）

商业综合体
1F-9F为商业综合体部分
其中1-5F为商业部分
6-9层为餐厅会所

地下车库

办公入口

商业次入口

酒店入口

商业主入口

休闲会所

餐饮空间

四层局部通高

观景面走道设小型茶座

缓冲空间进行布展

高折射率玻璃幕墙，开敞柔和

根据人群目的不同，空间流线可大致分为三类：

·旅游的客人
入口广场进入——休闲购物(1-3F)——参观用餐(4-5F)

·住店的旅客
酒店大堂办理手续(1F)——楼上餐厅用餐(6-7F)——
餐后会所休闲娱乐(10F)——到自己房间休息(26-38F)

·办公工作人员
早上来到办公楼(1F)——直接到办公区上班(11-24F)
——午餐来到员工餐厅(8F)——下班后会友（1-5F)

设计题目： 重峦 —— 高层绿色建筑课程设计
指导老师： 袁朝晖
学　　生： 姜熠、陈醒颖

8

Design topic: Range upon Range of Mountains—Course Design of High-rise Green Building

Instructor: Yuan Zhaohui

Students: Jiang Yi, Chen Xingying

● 设计说明

大桥的遮挡下，对于双塔来说，狭窄的景观视线并不能满足其对于滨江景观的需求；考虑到高层建筑对于城市环境的影响，我们希望在借用现有景观的同时，整个建筑对城市环境不造成坏的影响，呈现出良好的景观，然后使建筑内部形成良好的景观，最后形成"远景—中景—近景"的景观模式，使外向型景观与内向型景观实现巧妙的结合。

Design notes

Under the shelter of the bridge,the limited landscape of the twin towers can not meet people's needs for riverside landscape. Considering the influence of the high-rise buildings on the urban environment,we hope that the whole building presents a good landscape in the urban environment and forms a good landscape inside the building in the use of the existing landscape,and finally form the pattern of distant vision to middle vision to close vision ,so that the exterior landscape and the interior landscape can reach clever combination.

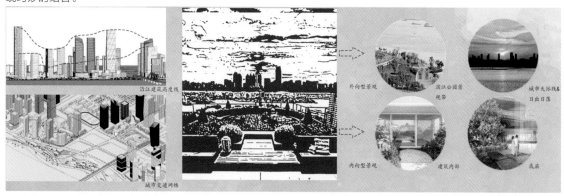

形体生成

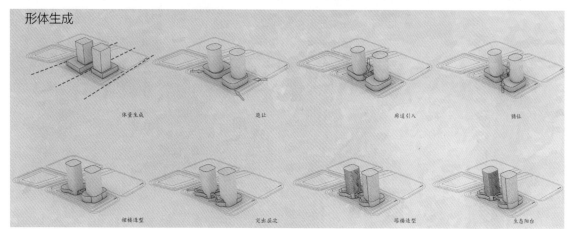

9

设计题目： Scherk's Plaza—— 建筑设计 V 高层建筑设计
指导老师： 齐靖
学　　生： 陈大鹏、汤晟晖

Design topic: Scherk's Plaza—Architectural Design V: High-rise Building Design

Instructor: Qi Jing

Students: Chen Dapeng，Tang Chenghui

● 设计说明

新的商业发展模式应该注重加强垂直化商业的沟通与联系，促进各个空间的互通，使得其在获得大众资本需求的同时，也与其他个体取得交流，形成一定的纽带关系，而非完全的竞争关系。本方案采取路网折叠的设计方法，将平面交通叠加至垂直向，使得不同高度的人群与商业有更加紧密的联系，结合地形建立起多个缓坡与垂直交通，形成完备的 TOD 模式。同时在 11~13 层建立起空中连廊，形成穿插交通体系，达到增加联系、空间互通、获得交流的目的。

Design notes

The new business development model should pay attention to strengthening the communication and connection of vertical commerce and promoting the interconnection of various spaces,so that it can not only meet the public capital's need,but also communicate with other individuals,and form a certain bond relationship rather than a complete competitive relationship.In this scheme, the design method of road network folding is adopted,and the plane traffic is superimposed to the vertical direction,so that people at different heights have closer contact with commerce. Combined with the terrain, multiple gentle slopes and vertical traffic are established to form a complete TOD mode.At the same time, an air corridor is established between floors 11 and 13 to form an interspersed traffic system,so as to increase contact,share space and obtain communication.

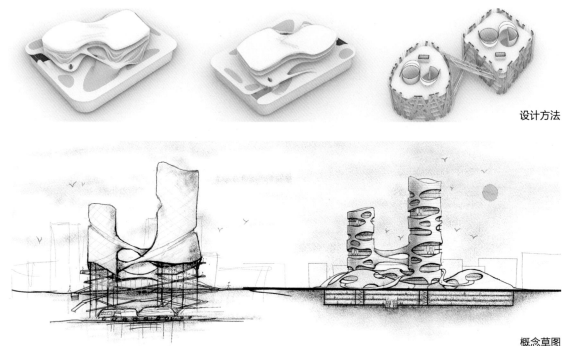

设计方法

概念草图

10

设计题目： 叠峦 —— 湖南大学第二体育馆设计
指导老师： 袁朝晖、邓广
学　　生： 乔畅、曾亦罗

Design topic: Overlapping Hills—The Design of the Second Gymnasium of Hunan University

Instructors: Yuan Zhaohui, Deng Guang

Students: Qiao Chang, Zeng Yiluo

● **设计说明**

设计从周边环境出发，首先考虑东西两侧的山体，四个抛物直纹曲面组合形成的结构单体，可以反映场地东西侧的山形意象；场地周边建筑群大多有韵律感地重复某个单元，将山形特征和周边建筑特征组合，即得到了建筑形态。主体采用格构式交叉钢架结构，结合索膜，实现大跨空间，形成丰富的建筑形态。

Design notes

Starting from the surrounding environment,the design first considers the mountains on the east and west sides and structural monomer formed by four parabolic ruled surfaces,which can reflect the mountain trend on the east and west sides of the site.Most of the surrounding buildings repeat a certain unit rhythmically, and combine the characteristics of the mountain shape with those of the surrounding buildings to obtain the architectural form.The main body adopts lattice type cross steel frame structure and combine with cable membrane to achieve a large span space and form abundant architectural forms.

场地设计

方案生成

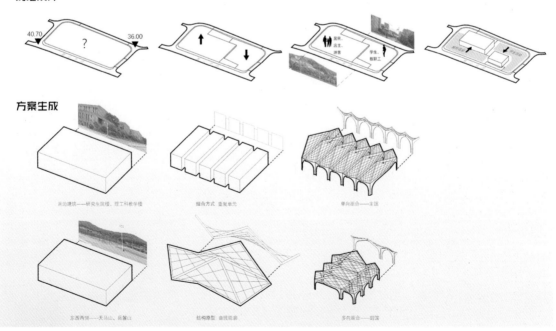

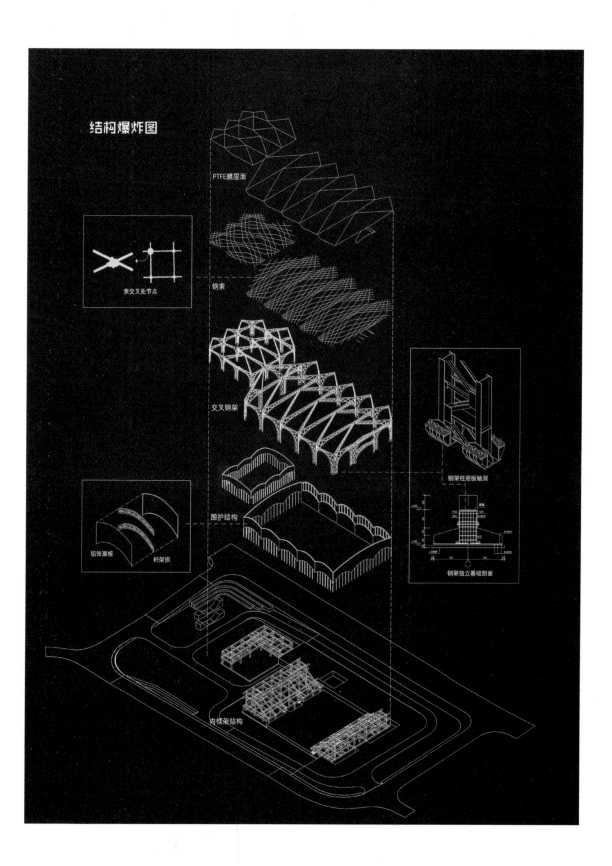

结构爆炸图

PTFE膜屋面

索交叉处节点

钢索

交叉钢架

钢架柱底板轴测

围护结构

铝饰面板   桁架拱

钢架独立基础剖面

内框架结构

采光分析 ——计算主席组内场

ECOTECT模型　　　自然采光系数　　　自然采光照度

一般幕材料

使用透光膜

表4.0.14 体育建筑的采光标准值

分析

结论

辐射分析 ——计算西立面辐射

ECOTECT模型　　　夏季日均辐射

方案一
不挑檐

方案二
挑檐

方案三
挑檐上倾开窗

方案四
挑檐下倾开窗

分析

结论

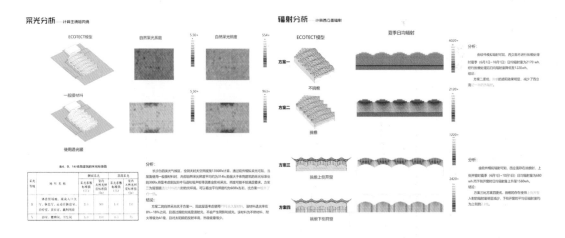

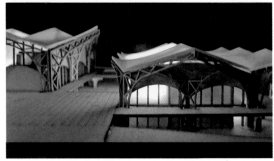

11

设计题目： 鹰起山中 —— 大跨度建筑课程设计
指导老师： 陈晓明、邓广
学　　生： 徐子牧、温俊嘉

Design topic: Eagle From the Mountain—Large Span Architecture Course Design

Instructors: Chen Xiaoming, Deng Guang

Students: Xu Zimu, Wen Junjia

● **设计说明**

本次设计基于以下三个矛盾点：首先，基地东西两侧分别有山，处在具有纪念意义的校园建筑环境中，如何处理同山以及校园建筑的关系，凸显大跨度体育建筑的性格，是本次方案需要解决的主要矛盾；同时，基地两侧临城市道路以及内部道路，设计中需要考虑如何消解大跨建筑的巨大体量对两侧道路的压迫，来达到校园立面的统一。

Design notes

The design is based on the following two problems:first of all,there are mountains on the east and west sides of the base,which is in a commemorative campus building environment,so how to highlight the character of large-span sports buildings is the main problem to be solved in this scheme through dealing with the relationship between the mountains and campus buildings;second,there are urban roads and internal roads on both sides of the base.Therefore,we need to consider how to eliminate the pressure of the huge volume of the large-span buildings on the roads in the design,so as to achieve the unification of the campus facade.

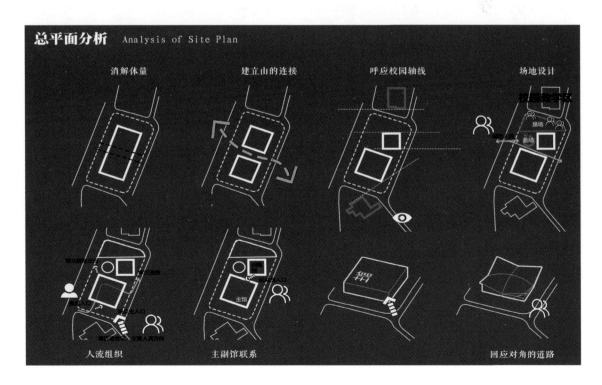

总平面分析 Analysis of Site Plan

消解体量　　建立山的连接　　呼应校园轴线　　场地设计

人流组织　　主副馆联系　　回应对角的道路

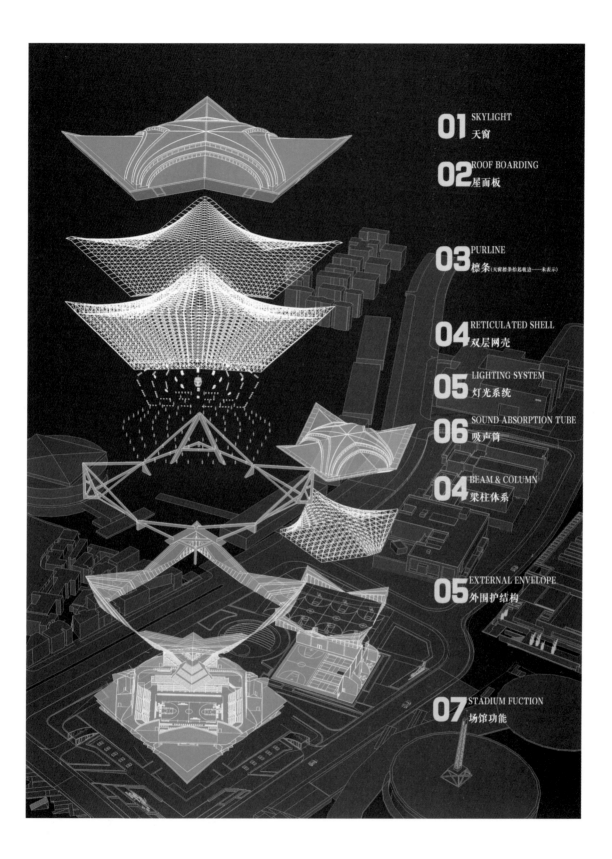

01 SKYLIGHT
天窗

02 ROOF BOARDING
屋面板

03 PURLINE
檩条 (天窗檩条抬起收边——未表示)

04 RETICULATED SHELL
双层网壳

05 LIGHTING SYSTEM
灯光系统

06 SOUND ABSORPTION TUBE
吸声筒

04 BEAM & COLUMN
梁柱体系

05 EXTERNAL ENVELOPE
外围护结构

07 STADIUM FUCTION
场馆功能

露天篮球场　活动广场　露天剧场　地下车库　生态跑道　室外健身广场　运动雕塑　比赛场馆　室外乒乓球场　集散广场

187

设计题目： 檐下 —— 湖南大学第二体育馆设计
指导老师： 卢健松、邓广
学　　生： 李佳、姚怡林

12

Design topic: Under the Eaves—the Design of the Second Gymnasium of Hunan University

Instructors: Lu Jiansong, Deng Guang

Students: Li Jia, Yao Yilin

总平面图 Site-plan

基地现状 Base Situation

场地分析 Site Analysis

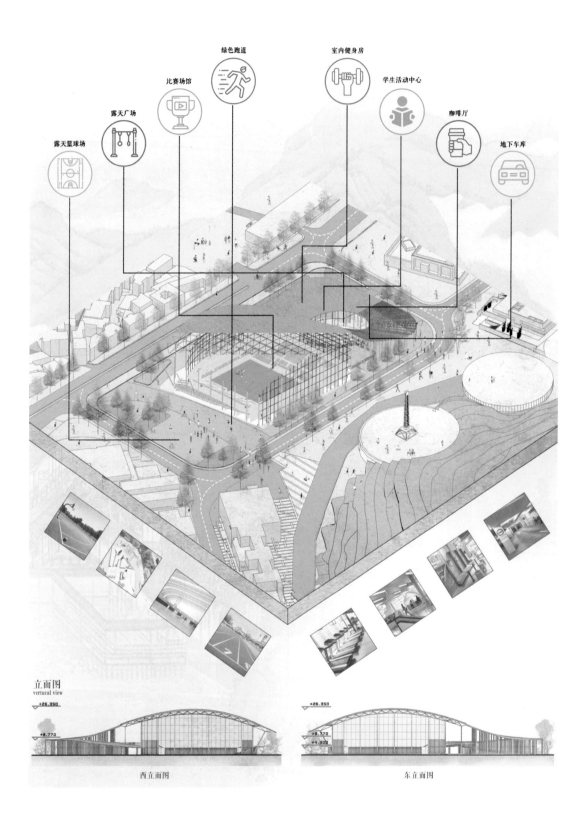

露天篮球场

露天广场

比赛场馆

绿色跑道

室内健身房

学生活动中心

咖啡厅

地下车库

立面图
vertucal view

+26.350

+8.770

西立面图

+26.350

+8.770
+4.300

东立面图

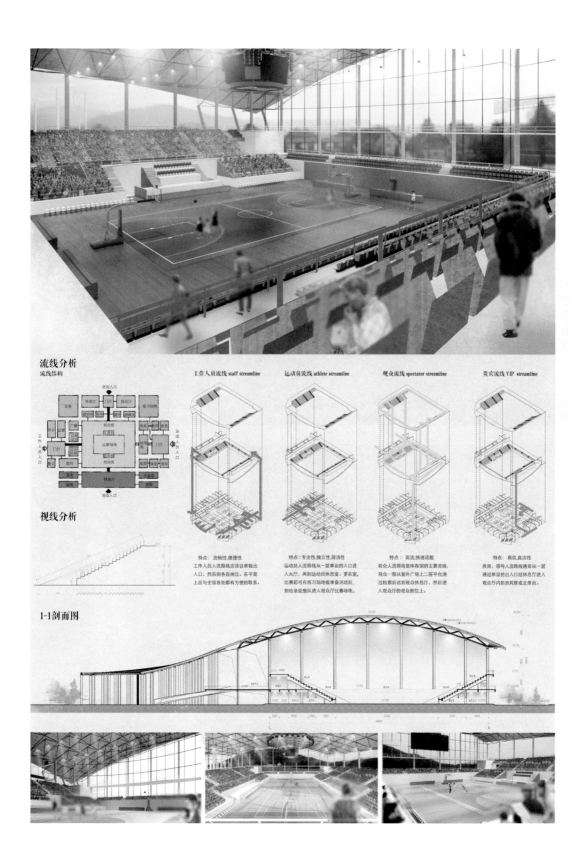

流线分析

流线结构

贵宾入口

工作人员流线 staff streamline

运动员流线 athlete streamline

观众流线 spectator streamline

贵宾流线 VIP streamline

视线分析

特点：流畅性、便捷性
工作人员入流路线应设设单独出
入口，然后到各自岗位，在平面
上应与全馆各处都有便的联系。

特点：专业性、独立性、清洁性
运动员入流路线从一层单设的入口进
入大厅，再到运动员休息室、更衣室，
比赛前可在练习场地做准备活动后，
到检录处整队进入观众厅比赛场地。

特点：简流、快速疏散
观众人流路线是体育馆的主要流线，
观众一般从室外广场上二层平台通
过检票后达到观众休息厅，然后进
入观众厅的观众席位上。

特点：高效、直达性
贵宾、领导人流路线通常从一层
通过单设的出入口经休息厅进入
观众厅内的贵宾席或主席台。

1-1剖面图

设计题目： 壑谷 —— 湖南大学第二体育馆设计
指导老师： 刘尔希、邓广
学　　生： 任意、冉富雅

13

Design topic: The Valley—the Design of the Second Gymnasium of Hunan University

Instructors: Liu Erxi, Deng Guang

Students: Ren Yi, Ran Fuya

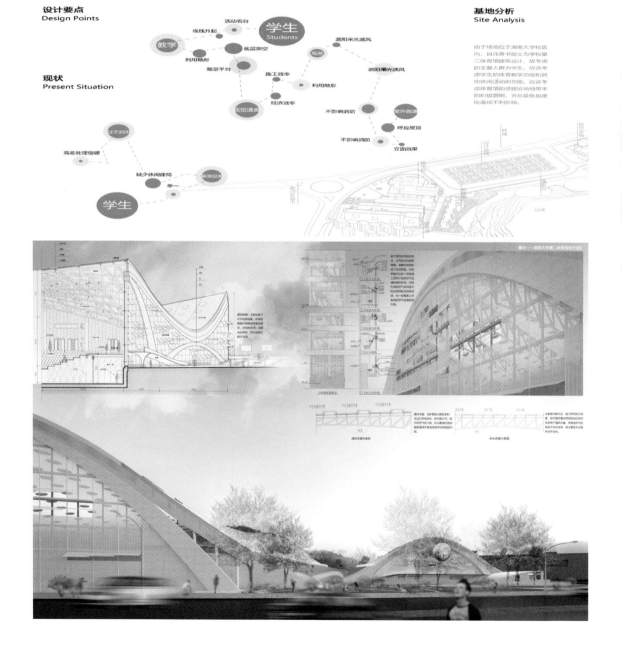

概念提取
Concept extraction

多种功能	场地高差	使用人群	通透性	绿色可持续
？	？	？	？	？

不仅提供了室内外教学的功能，也从学生角度出发，尽可能丰富大家的活动和感受。

巧妙运用场地1m的高差，需大程度地减少开挖水平，经济效益最大化。

使用人群大多数确定为学生，所以确定了建筑的主入口在面向研究生院路方向。

靠场地无建筑遮挡，放观线与场地较小幅度地做环场地，保留视觉通透。

东西向长边，故建筑适应场地东西向长边时应该重点考虑遮阳通风的处理，以及自然光的采集。

形态提取
Morphological extraction

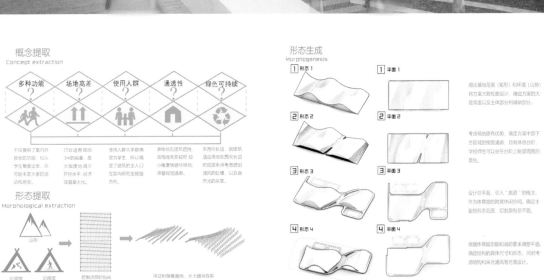

山形   控制点间时抬向   保证折架是直线，大大提高效率

比赛馆   训练馆

形态生成
Morphogenesis

1 形态 1	1 平面 1

顺应基地范围（矩形）和环境（山势），将方案大致轮廓设计，确定方案的大致高度以及主体部分和辅助部分。

2 形态 2	2 平面 2

考虑场地地形有优势，确定方案中部下方区域的视觉通畅，且有休息台阶，学校师生可以坐在台阶上眺望周围的景色。

3 形态 3	3 平面 3

设计总平面，引入"廊道"的概念，作为体育馆的附属休闲空间，确定主题馆形态范围，切割数据有总平面。

4 形态 4	4 平面 4

根据体育馆功能和消防要求调整平面，确定结构的具体尺寸和形态，同时考虑结构和采光通风等方面设计。

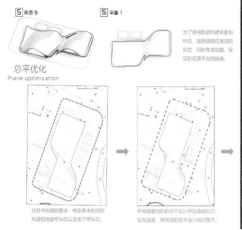

S 形态 5   **S** 平面 1

总平优化
Plane optimization

为了使得跑道和建筑更加呼应，故跑道屋顶的形式，同时考虑功能，保证和观景平台的连接。

任务书和消防要求，确定基本的消防环道和地面停车位以及地下停车位。

中间观景台阶的总平设计呼应跑道形式，较为活泼，两端的总平设计相对整齐。

经济技术指标

序号	名称	单位	数量	备注
一、用地规模	建筑用地面积	m²	21245.0	
二、建筑规模	总建筑面积	m²	15954.8	
	其中 地上建筑面积	m²	3303.0	计容面积
	地下一层建筑面积	m²	9314.8	计容面积
	机动车库	m²	3337.0	不计容面积
三、建筑基底面积	建筑基底面积	m²	9023.8	
	道路广场面积	m²	10062.8	
	绿地面积	m²	2158.4	
四、相关指标	绿地率	/	10.2%	
	建筑密度	/	42.5%	
	容积率	/	0.69	
	建筑高度	m	21.5	
其中 机动车停车位		个	64	
	地上	个	16	
	地下	个	48	

设计说明

1. 建筑名称：湖南大学第二体育馆
2. 本项目位于湖南大学南校区天马山与麓山南路之间的用地。
3. 建筑用地面积：21245.0㎡ 总建筑面积：15954.8㎡ 建筑高度：21.5m
   建筑等级：乙级 耐火等级：二级 结构类型：框架及大跨度钢桁架结构体系
4. 本工程设计绝对标高±0.000相当于绝对标高39.70m，本项采用大地坐标系：高程系统为黄海高程系统。高程、尺寸、距离以"m"计
5. 建筑四周建筑环形消防车道通。

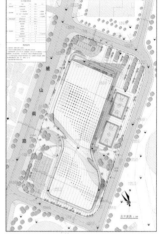

总平面图 1:500

室内透视
Interior Perspective

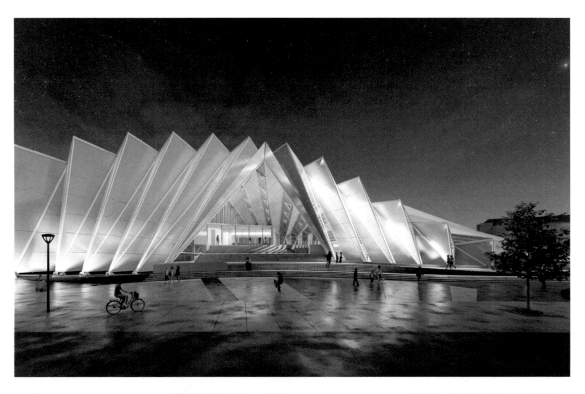

14

设计题目： 折峦见景 —— 湖南大学第二体育馆设计
指导老师： 严湘琦、邓广
学　　生： 周枫、苏雯玲

Design topic: See the Scenery through the Mountains—the Design of the Second Gymnasium of Hunan University

Instructors: Yan Xiangqi,Deng Guang

Students: Zhou Feng,Su Wenling

● 基地分析

基地位于湖南大学内，西侧为城市道路，车流量较大，使用者主要为学生、市民、游客；东侧为校内道路，车流量小，主要使用者为学生；南北侧均为教学建筑，西侧为沿街店铺及居民住宅，与南北侧教学建筑共同形成校内区域与校外区域的分界线。且场地位于两山之间，两山分别为岳麓山与天马山，为了防止视线遮挡，对建筑有限高要求。

Base analysis

The base is located in Hunan University.The west side of the base is the urban road with a large traffic flow which are mainly used by students,citizens and tourists. There are education buildings in the north and south side of the base, and shops and residential houses along the street in the west form the boundary between the area inside the school and the area outside the school.As the site is located between the two mountains, Yuelu Mountain and Tianma mountain,in order to prevent the line of sight from blocking,the construction are limited to height requirements.

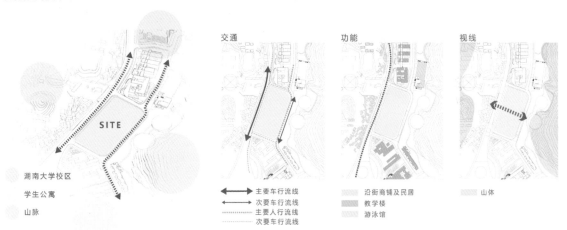

交通　　　　　功能　　　　　视线

湖南大学校区
学生公寓
山脉

→ 主要车行流线
→ 次要车行流线
⋯⋯ 主要人行流线
⋯⋯ 次要车行流线

▨ 沿街商铺及民居
▨ 教学楼
▨ 游泳馆

▨ 山体

● 概念生成

方案主要基于以下考虑：（1）对两侧山脉景色的呼应；（2）基地在校园中分隔校园内外道路的地理位置；（3）校园体育馆活力的展示。因此，方案选择南北连续的体量保持校园建筑的整体布局，使校园内外道路有一定的独立性。结合折板结构，使体育馆形态顺应两侧山势，将立面"掀开"展示内部活动场景，以吸引人群的进入，并塑造了"景中景"的沿街立面效果，山景前将体育馆与山势协调成为"物景"，体育馆立面打开，展示校园富有活力的学生活动景色，形成"人景"，形成沿街立面有层次的景象效果。

Concept generation

The plan is mainly based on the following considerations: 1.The echo to the scenery of the mountains on both sides. 2.The geographical location of the base is in the campus that separates the roads inside and outside the campus.3.Display of vitality of campus gymnasium.Therefore, the plan selects the continuous volume from north to south to maintain the overall layout of the campus buildings,so that the roads inside and outside the campus have a certain independence.Combined with the folded plate structure, the shape of the gymnasium conforms to the mountains on both sides and the facade is "lifted" to show the internal activity scene, so as to attract people to get in,and shape the street facade effect of "scenery in the scene". The mountain view harmonizes the gymnasium with the mountain into a "physical view" and the facade of the gymnasium is opened to show the dynamic student activity scenery of the campus, forming a "human view", which together presents a hierarchical scene effect along the street facade.

15

设计题目：假如今日没有剧目上演 —— 校园文化艺术中心设计
指导老师：齐靖、邓广
学　　生：朱芸桦、袁振香

Design topic: If There is No Play Today—Design of Campus Culture and Art Center

Instructors: Qi Jing，Deng Guang

Students: Zhu Yunhua，Yuan Zhenxiang

● **设计说明**

强调建筑的日常性，通过内凹的曲线产生室内与室外模糊的柔和空间，同样的母题形成的"洞穴"作为吸纳学生活动的场所，也成为面向周围山形的观景空间，屋顶形态顺应"风"吹过的曲线，撕开几处屋面使自然风能贯通屋顶和下部的洞穴，并引入自然光。建筑内部通过观景腔体组织交通并划分空间，试图连接场地西面熙熙攘攘的麓山南路和场地东面安静的天马山与学生路径这两种截然不同的场所氛围。由此，纳入时间，纳入自然，纳入声音，纳入行人，通过纳入式建筑的手法打破传统观演建筑大尺度的压迫感，使其成为自然的风雨、大地里的鸟虫和学生日常生活共同发声的腔体，即使今日没有剧目上演。

Design notes

The daily nature of architecture is extremely important.Through the concave curve, the indoor and outdoor fuzzy soft space is generated,and the formed "cave", as a place to absorb students' activities, has also become a viewing space facing the surrounding mountains.The shape of the roof conforms to the curve of the wind,tearing several roofs to make the natural wind run through the roof and the lower cave,and introduce natural light.The interior of the building is divided into spaces through the viewing cavity,trying to connect the two distinct atmosphere of Lushan South Road in the west and the quiet student path in the east.Therefore,through the method of inclusive architecture:incorporating time, nature, sound and pedestrians to break the oppression of traditional viewing performance architecture.It is also a cavity for natural wind and rain,birds and insects in the earth and students' daily life,even if there is no play today.

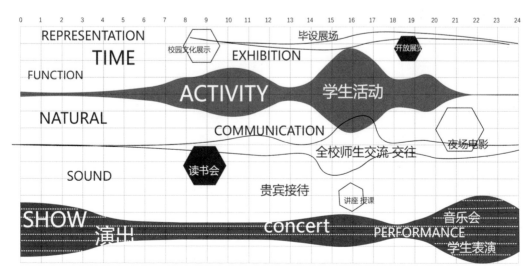

失联

无变化

分隔

割裂

场地分析:

建筑风格统一但趋于严肃和单一。整条学生路径缺乏"可停留空间"。考虑长沙冬日潮湿多雨,夏日干燥炎热的气候,我们认为在连接校区与学生宿舍的路径中点应该有一座建筑接纳往来学生,产生荫蔽或者其他多样的光线,联系四周的环境并引入自然风,产生新的可自由通过的路径,打破现有的"垂直墙体的房子、宽宽、无变化的道路、被拉远的山体"组合的割裂化的场地环境。

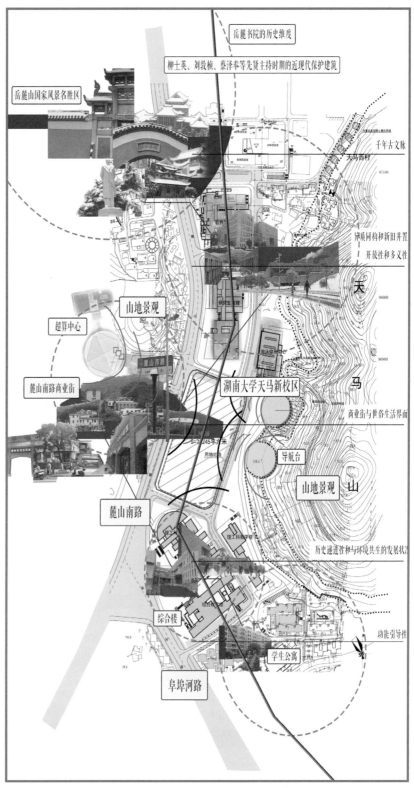

岳麓书院的历史维度

柳士英、刘敦桢、蔡泽奉等先贤主持时期的近现代保护建筑

岳麓山国家风景名胜区

千年古文脉

天马西村

异质同构和新旧并置

开放性和多义性

山地景观

超算中心

麓山南路商业街

湖南大学天马新校区

商业街与世俗生活界面

导航台

山地景观

天 马 山

麓山南路

历史递进延种与环境共生的发展状态

综合楼

功能引导性

学生公寓

阜埠河路

200

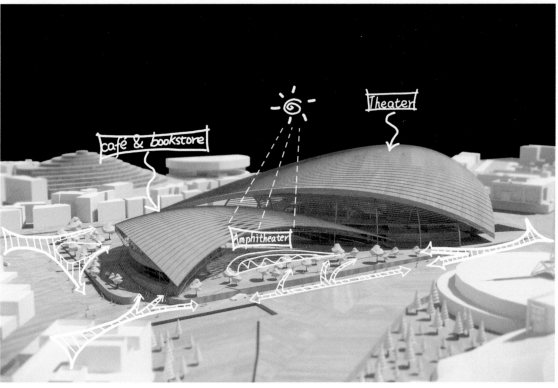

café & bookstore

Theater

Amphitheater

设计题目： HUN·剧院十二时辰 —— 湖南大学校园文化艺术中心设计
指导老师： 齐靖、邓广
学　　生： 陈颂、赫洪

16

Design topic: HNU·Twelve Two-Hour Periods of the Theater—Design of Campus Culture and Art Center of Hunan University

Instructors: Qi Jing，Deng Guang

Students: Chen Song，He Hong

● 设计说明

基地位于湖南大学校区内部，校区内人流复杂且密集。基地西侧是一些饮食类小商铺，东侧是校内道路，比较安静整洁。功能融合思考，项目作为校园内部文化艺术中心的重要节点，其核心的影剧院功能与学生日常生活的关联值得思考。文化艺术中心与影剧院对应的不同功能和人群产生化学效应，即不同时间节点场地内的人流路线及使用导向能够成为建筑形体生成的引发器，"十二时辰"这一概念应运而生。

Design notes

The base is located inside the campus of Hunan University, where the flow of people is complex and dense.There are some small catering shops in the west side of the base, and school roads in the east side are relatively quiet and tidy. As an important node of the Campus Culture and Art Center, the connection between its core theater function and students' daily life is worth thinking about. The different functions and groups corresponding to the Campus Culture and Art Center and the theater produce chemical effects that the pedestrian flow route and use guidance in the site at different time can become the initiator of architectural form generation. Therefore, the concept of "twelve two-hour periods" came into being.

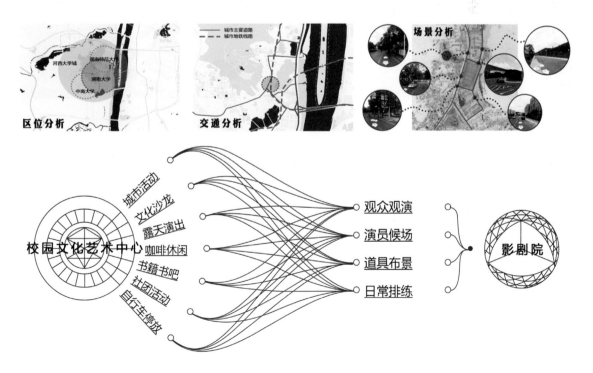

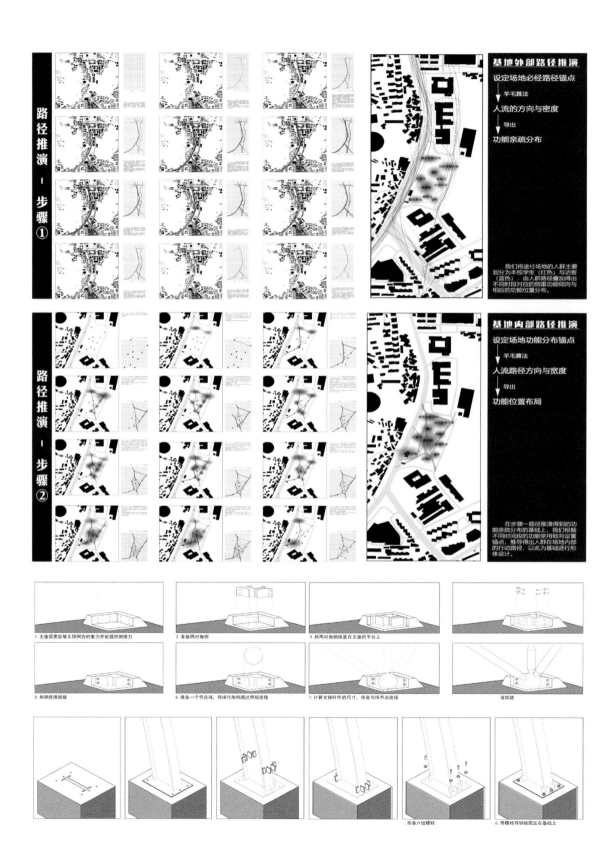

路径推演 - 步骤①

基地外部路径推演

设定场地必经路径锚点
↓
羊毛算法
↓
人流的方向与密度
↓
导出
↓
功能亲疏分布

我们将途经场地的人群主要划分为本校学生（红色）与访客（蓝色），由人群路径叠加得出不同时段对应的侧重功能倾向与相应的功能位置分布。

路径推演 - 步骤②

基地内部路径推演

设定场地功能分布锚点
↓
羊毛算法
↓
人流路径方向与宽度
↓
导出
↓
功能位置布局

在步骤一路径推演得到的功能亲疏分布的基础上，我们根据不同时间段的功能使用倾向设置锚点，推导得出人群在场地内部的行动路径，以此为基础进行形体设计。

1 支座需要旋转支撑网壳的重力并能提供侧推力

2 准备两对角钢

3 将两对角钢放置在支座的平台上

5 角钢连接就绪

6 准备一个节点球，将球与角钢通过焊接连接

7 计算支撑杆件的尺寸，预备与球节点连接

焊接就绪

准备六组螺栓

6 用螺栓将钢板固定在基础上

204

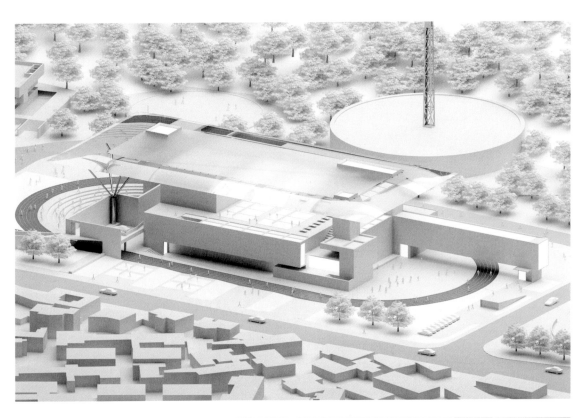

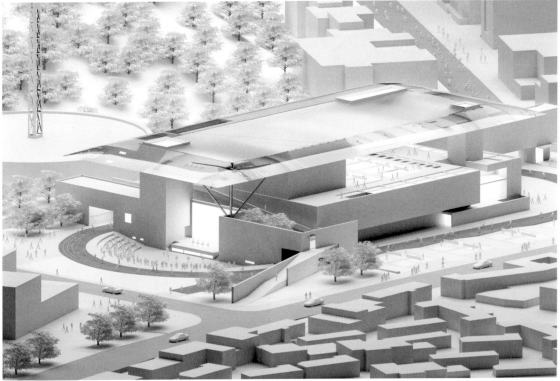

設計題目：文体方舟 —— 湖南大学校园文化艺术中心设计
指导老师：宋明星、邓广
学　　生：孙凡清、邱子倍

17

Design topic: Culture and Sports Ark—Design of Campus Culture and Art Center of Hunan University

Instructors: Song Mingxing，Deng Guang

Students: Sun Fanqing，Qiu Zibei

● 设计说明

因湖南大学用地紧缺，运动空间同样不足，故设计者保留场地原有运动场，文艺与体育相结合，并采用体块组合设计手法延续场地文脉。"敞开的"外壳增加建筑内文艺活动与外部运动之间的互动，创造非日常/日常事件交集。薄顶下的半户外运动空间在夏季成为校园景观凉厅。镜面顶连接四方景观/事件，提升场地活力。建筑体块由"开口筒体"逻辑支配，筒体向校园"打开"，营造日常氛围；筒体向城市封闭，创造有纪念性的建筑体量。

Design notes

Due to the shortage of lands in Hunan University and the lack of sports space,the designer retains the original sports ground,combines literature and art with sports, and uses the block combination design method to continue the context of the venue.The "opened" shell increases the interaction between literary and artistic activities in the building and external sports activities and creates the intersection of non daily or daily events.The semi-outdoor sports space under the thin dome becomes a campus landscape cool hall in summer.The mirror links four landscapes and events to enhance the vitality of the site.The building block is dominated by the logic of "open cylinder",which is "opened" to the campus to create a daily atmosphere and closed to the city to create a memorial building volume.

1.保留原有跑道，将文艺中心集中在跑道内测，实现运动与文艺的复合，增加场地的活力。

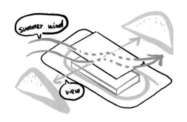

2.掀开屋顶，形成景观通廊，同时有利于夏季通风乘凉，为学生提供可活动的公共凉厅。

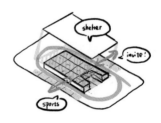

3.屋顶下部灰空间设半室外运动区，提供舒适运动环境，与场地两侧运动区连接贯通。

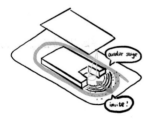

4.建筑北侧设室外舞台与跑道耦合，实现看表演和运动事件的交叉。

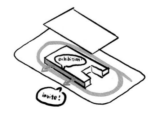

5.建筑东侧设流动展厅与跑道耦合，实现看展和运动事件的交叉。

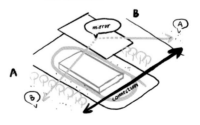

6.建筑天花设镜面不锈钢反射场地两侧景色，实现场地的视线连接。

设计题目： 立石馆集合群
指导老师： 刘尔希
学 生： 何逸伦、梁祝

18

Design topic: Steles Museum Group

Instructor: Liu Erxi

Students: He Yilun，Liang Zhu

● **视图建筑**

1931~1932 年，巴拉甘作品最早的出版物中常常是朦胧的深讷色色调和相当缺乏想象力的构图，现在取而代之的是充满戏剧性和活力的清晰黑白图像。当巴拉甘搬到墨西哥城后，他开始对欧洲的功能主义建筑非常感兴趣(1935)，例如在墨西哥城 Parque 大道（1936 年）的复式公寓的四张图片中，清晰的光线穿透下来形成阴影，墙内功能和位置显示得并不清楚，朴素的浅色墙壁以惊人的切角分隔子结构框架，大地或天空的一角或落或升，窗框的影子斜斜地映在一面墙上，而另一种烟囱状的装置则像长腿的鸟在电线上飞起。这些作为整体拍摄的图像十分吸引人，暗示着连接内外空间运动的戏剧性。然而它们也是碎片，对于运动将如何进行，以及各个部分将如何组合成一个整体，提供的信息很少。照片旁边的平面对于理解它们几乎没有指导价值，其实这些照片也无助于阅读平面。他也发表了一些他早期的瓜达拉哈拉房子的局部照片，这类带有一点误导性的图片忽视了原作中的木制家具、拱形开口和历史主义的细节装饰，而是重点展示它们的简洁、立方体特质和类似柯布西耶的现代形式下的光影。从那时开始，这些图像似乎是为了研究或展示图像是如何操纵建筑形式、激发读者想象力而设计的。

● **View Building**

From 1931 to 1932, the earliest publications of Barragan's works were often full of hazy dark tones and rather unimaginative compositions, but now they are replaced by clear black and white images which are dramatic and full of vitality. When Barragan moved to Mexico City, he became very interested in European functionalist architecture (1935). For example, in the four pictures of duplex apartments on Parque Avenue (1936) in Mexico City, clear light penetrated glass and shadows are formed. The functions and positions of the walls are not clearly displayed. The plain light-colored walls separated the sub-structure frames with amazing cut corners. With the rise and fall of the sun, the shadow of the window frame is diagonally reflected on one wall and another chimney-like device flies up on the wire like a long-legged bird. These images taken as a whole are very attractive,suggesting the drama that connects the movement of space inside and outside. However, they are also fragments. There is little information on how the movement will proceed and how the various parts will be combined into a whole. The planes next to the photos have little guiding value of helping us understand them. In fact,these photos are not helpful for reading planes. He also published some partial photos of his early Guadalajara house. This kind of misleading pictures ignores the wooden furniture, arched openings and historicism details in the original work. The focus is to show their simplicity, cubic characteristics and the light and shadow in a modern form similar to Le Corbusier. Since then, these images seemed to be designed to study or show how images manipulate architectural forms and stimulate the imagination of readers.

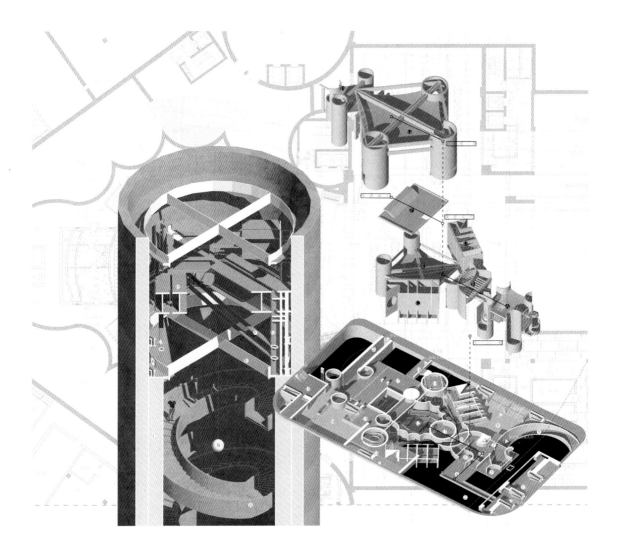

建筑划分部分

A. 门厅建筑　B. 过厅建筑　C. 中小型演艺综合体建筑
D. 公众商业服务建筑　E. 设备夹层　F. 小音乐厅建筑
G. 主剧场建筑（屋顶部分）H. 芭蕾舞排演厅建筑

功能布置部分

1. 入口平台　2. 静水池　3. 室外平台　4. 入口平台　5. 上
二层　6. 门厅休息　7. 去往公众广场　8. 广场　9. 地库坡
道　10. 过厅　11. 主交通（服务）塔　12. 观众厅　13. 办
公塔　14. 展厅集散平台　15. 舞台上层广场　16. 后广场
17. 舞台顶部广场　18. 演艺人员入口　19. 后部平台
20. 布景车电梯

Building Division

A.Foyer building　B.Hall building　C.Small and medium-sized
performing arts complex building　D.Public commercial service
building　E.Equipment mezzanine　F.Small concert hall building
G.Main theater building (roof part)　H.Ballet rehearsal hall building

Function Layout Part

1. Entrance platform　2. Silent pool　3. Outdoor platform
4. Entrance platform　5. Upper second floor　6. Foyer rest
7. Go to the public square　8. Square　9. Basement ramp
10. Passing hall　11. Main traffic (service) tower 12. Audience hall
13. Office tower　14. Exhibition hall distribution platform
15. Stage upper square　16. Rear square　17. Square at the top of
the stage　18. Entrance for entertainers　19. Rear platform
20. Scenery car elevator

设计题目： 山语间 —— 校园文化艺术中心课程设计
指导老师： 宋明星、胡娴
学　　生： 刘盛华、黄钊

19

Design topic: Shan Yujian—Curriculum Design of Campus Culture and Art Center

Instructors: Song Mingxing, Hu Xian

Students: Liu Shenghua, Huang Zhao

● **场地关系**

根据场地周边建筑及道路关系确定轴线布局，并根据基地自身需承担的穿行界面的特点形成主要布局，采用羊毛算法迭代优化路径，形成场地主要步行流线。

Site relationship

The axis layout is determined according to the surrounding buildings and road relations of the site, and the main layout is formed according to the characteristics of the crossing interface that the base needs to bear, and the wool algorithm is used to iteratively optimize the path to form the main pedestrian flow line of the site.

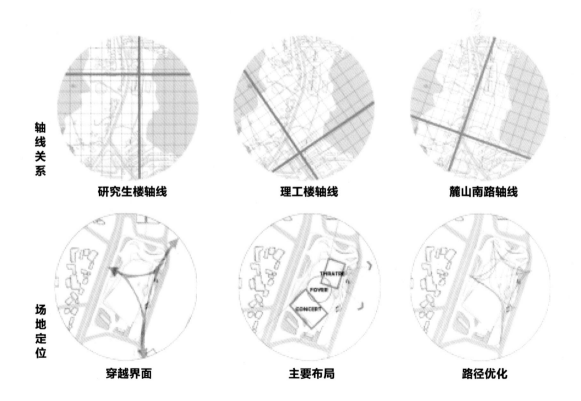

轴线关系

研究生楼轴线　　　　理工楼轴线　　　　麓山南路轴线

场地定位

穿越界面　　　　主要布局　　　　路径优化

● 形体生成　Shape generation

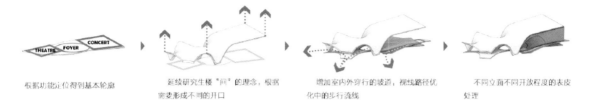

根据功能定位得到基本轮廓

延续研究生楼"间"的理念,根据需要形成不同的开口

增加室内外穿行的坡道,视线路径优化中的步行流线

不同立面不同开放程度的表皮处理

● 功能分区及流线分析　Functional zoning and streamline analysis

利用顶地势有高有低不同人群的分流，学生走坡地缓坡一侧进入，也一层作为学生的入口门厅，将进入的人引导向上，一层主要为学生可做活动场地，并承担对外的展示作用；二层为观众入口，通过台下走坡的观众人群与坡地楼梯间人群形成交流；三层且备据对外景观资源，为观种人群共同使用、交汇的场所。

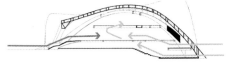

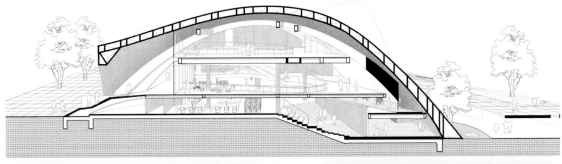

● 屋顶受力分析　Stress analysis of roof

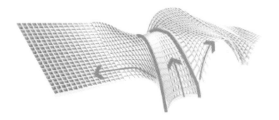

对屋顶形态进行力学模拟，发现网壳起拱最高处所受应力最大，且本身主要承受自重压力，故将网壳杆件由下往上进行截面的缩小以减小自重；另外，与拱衔接的区域应力相对集中，将与拱连接的杆件强化。

According to the mechanical simulation of the roof shape, we found that the highest part of the arched reticulated shell is subjected to the greatest stress and the main body is subjected to the pressure of its own weight.Therefore,the cross-section of the reticulated shell members is reduced from bottom to top to reduce the self-weight.In addition,the pressure in the area connected with the arch is relatively concentrated, so as to strengthen the rod piece which is connected with the arch.

城市设计

Urban design

城市设计课程介绍
Course Introduction of Urban Design

教师团队
Teacher team

叶强
Ye Qiang

蒋甦琦
Jiang Suqi

姜敏
Jiang Min

陈煊
Chen Xuan

向辉
Xiang Hui

彭科
Peng Ke

许昊皓
Xu Haohao

孙亮
Sun Liang

课程介绍
Course introduction

一、课程简介

城市设计课程，是"以现实问题为导向，以空间响应为手段"，培养学生运用空间认知解读、空间优化设计相关知识和技能，分析、解决城市现实问题的能力。本课程聚焦从城市研究中发现问题、分析方法和设计策略等内容，这既是一个对城市中观和微观尺度下的整体物质环境的塑造过程，也需要关注城市的经济、文化、政治以及人们的行为活动等非物质因素对这个过程的影响和作用。

课程基于城市特定区域的城市历史文化经纬，通过城市设计的理论、方法和技术手段来分析、构想、调整、改造并优化多样类型的城市空间。课题鼓励同学敞开并创新城市特定地段改造保护的思路，认真调研，同时汲取国内外经验以及本组实践的优秀历史先例中所采用过的理论、方法与技术手段。课程希望学生通过学习，初步掌握基本的城市设计原理，并具有一定的城市问题的解决能力和扎实规范的设计表达技能。

城市设计课程为我院建筑学与城乡规划专业大四期间的专业核心课程，相关课程包为城市道路交通规划、城市社会学、城市社会调研等。

1.Course introduction

The course of urban design, based on the principle of "practical problems as the guide and spatial response as the means", aims to cultivate students' ability to analyze and solve urban practical problems by using knowledge and skills related to spatial cognitive interpretation and spatial optimization design. This course focuses on finding problems, analyzing methods and designing strategies from urban research, which is not only a process of shaping the overall material environment of the city at the medium and micro scales, but also needs to pay attention to the effect and role of non-material factors such as economy, culture, politics and people's behavior in this process.

Based on the urban historical and cultural longitude and latitude of specific urban areas, the course analyzes, conceives, adjusts, transforms and optimizes various types of urban space through the theories, methods and technical means of urban design. Projects in this course encourage students to open up and innovate ideas on the transformation and protection of specific urban areas, conduct careful research, and learn from domestic and foreign experience and the theories, methods and technical means adopted in the excellent historical precedents of certain group. The course hopes students can master the basic principles of urban design, and obtain certain ability to solve urban problems and solid and standardized skills of design expression.

The course of urban design is the core course of architecture and urban and rural planning major during the senior year of our college. The relevant courses include urban road traffic planning, urban sociology, urban social research, etc.

二、课程内容

课程教学目标

通过城市设计的教学，使学生进一步了解我国国土空间规划体系及城市建成环境体制的基本内容，掌握控制性详细规划（中观）尺度城市设计的工作内容和方法，能够正确运用相关理论知识、设计方法和各种技术手段，熟悉国家有关的技术规范标准，能够具备开展中观尺度城市设计工作和参与城市规划管理工作的基本知识和能力，理解城市空间特点和规划控制要求，并能在空间形态的设计与控制中予以回应。

通过本课程的教学，使学生具备以下能力：

1. 培养在多学科团队中工作的技能和建立在数字环境中工作的能力，完成的设计内容及深度、图纸表现均应达到规定要求。

2. 了解社会经济变化趋势和问题对城市空间环境要求的物理和感官指标，熟悉并能处理好城市设计所需的一系列空间属性：建筑形式，广场、绿地、环境小品等公共空间，景观，道路交通系统，停车位，密度，活动地点和强度等内容，强调获得记录和表达观察到的内容的技术，并增强以图形和书面形式传达城市设计思想的能力。

3. 培养城市设计师的全球视野和地方文化特征下的独立思维能力。

4. 培养学生的自主学习能力和创新能力，鼓励特定历史文化背景下的城市设计理念创新，提高对城市建筑群整体空间形态和城市空间环境设计的把握能力。

2.Course content

Teaching objectives

Through the teaching of urban design, students can further understand the basic contents of China's land and space planning system and urban built environment system, master the work contents and methods of regulatory detailed planning (meso) scale urban design, correctly use relevant theoretical knowledge, design methods and various technical means, and be familiar with relevant national technical specifications and standards, be able to have the basic knowledge and ability to carry out meso scale urban design and participate in urban planning management, understand urban spatial characteristics and planning control requirements, and respond in the design and control of spatial form.

Through the teaching of this course, students will acquire the following abilities:

2.1 Cultivate the skills of working in a multidisciplinary team and the ability to work in a digital environment. The completed design content, depth and drawing performance shall meet the specified requirements;

2.2 Understand the physical and sensory indicators required by socio-economic trends and problems for urban space environment, be familiar with and be able to deal with a series of spatial attributes required by urban design: architectural form, public space such as squares, green spaces and environmental sketches, landscape, road transportation system, parking space, density, activity location and intensity, etc., emphasize the acquisition of skill to record and express the observed content, and enhance the ability to convey urban design ideas in graphic and written forms;

2.3 Cultivate the independent thinking ability of urban designers under the global vision and local cultural characteristics;

2.4 Cultivate students' independent learning ability and innovation ability, encourage the innovation of urban design concept under the specific historical and cultural background, and improve their ability to grasp the overall spatial form of urban architectural complex and urban spatial environment design.

课题一

凤凰县老城区旧城更新城市设计
组织人：向辉、叶强

Topic 1: Urban Design of Phoenix County Old Town Old Town Updates

本次课程设计主题为凤凰县沱江镇凤凰古城核心保护区及周边地块的保护更新规划。

一、设计任务

1. 课程小组合作完成保护规划核心内容文本、图纸、说明书。
2. 学生独立完成地块更新规划控制导则。
3. 学生独立完成重要更新地段修建性详细规划。
4. 学生独立完成更新规划设计说明书。

二、设计目的

通过本次课程设计，增强学生对各类规范的熟悉程度，鼓励学生利用新技术，对历史文化名城进行分析。结合城市现状对历史街区的利用发展提出概念并进行规划设计，提高学生的设计水平和实践经验，鼓励学生立足居民及游客需求，将设计与社区各类人群的参与相结合，提高其设计服务意识，为以后步入行业和社会打下基础。

三、考核方式及要求

除保护规划内容部分采用小组合作方式进行外，本设计强调个人独立完成设计任务，强调概念表达的空间实现。最终成绩以个人完成工作的数量和质量作为评分标准，其中：小组工作量成绩占比不超过 30%，个人工作量占比不低于 70%。学生应在设计成果中明确表示出小组合作部分成果及个人完成内容；图纸绘制应符合《历史文化名城名镇名村保护规划编制要求》《城市规划编制审批办法实施细则》以及国家和地方相关规划标准，注重设计成果的规范化表达，重视图纸的工程性；毕业作品应结合任务场地实际情况，且有一定设计特色和思想。

四、成果内容要求

1. 保护规划核心内容文本、图纸、说明书（小组合作完成），图纸内容包括：
· 区位分析图
· 历史分析图
· 用地现状图
· 道路系统现状图
· 建筑质量现状图
· 建筑风貌现状图
· 建筑年代现状图
· 建筑高度现状图
· 历史遗存现状图
· 保护区划图
· 保护规划总图
· 用地规划图
· 道路系统规划图
· 高度控制规划图
· 开发强度控制规划图
· 片区更新方式示意图
· 建筑保护整治模式图
· 实施管控分类示意图

2. 更新规划、重点更新地段修建性详细规划、设计说明书（个人独立完成），图纸内容包括：
· 更新地块规划控制导则
· 重点更新地段规划平面图、立面图、鸟瞰图
· 其他用于表达设计意图的分析图

五、基地范围介绍

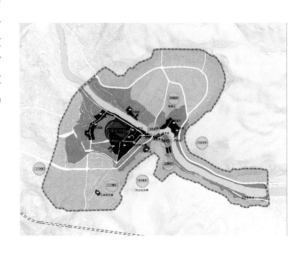

课题二

长沙市裕南街道东瓜山地块城市设计
组织人：陈煊

Topic 2: Urban Design of Changsha Yunan Street Donggua Mountain Plot

一、设计主题

此次课程培养学生对公共空间敏锐的观察能力、对社会文化空间公平客观的支持态度，并能够运用丰富的专业知识和手段分析城市问题，建立和培养"以人为本"的设计理念和方法。鼓励学生观察、体验、分析城市空间形态问题、空间产品的社会经济问题及历史、人文等问题，从而以空间规划的专业基础和引导"城市人"的合理行为作为基本手段，发掘城市文化和社会发展的背景，提出形态设计方案，继而丰富文化、和谐发展，并以全面、系统的专业素质去处理城市设计问题。

二、设计内容

1. 概况简介

东瓜山片区隶属长沙市天心区裕南街街道，位于书院路以东，白沙路以南，解放四村以西，南湖路以北的东瓜山南侧。社区面积约 0.2km²，辖区内有 62 栋楼房，辖区内无单位，现有居民 2587 户、5437 人，其中流动人口 560 人。裕南街社区坐落于办事处的南部，东起东电社区，西至宝塔山社区，南起石子冲社区，北至白沙南路，辖区面积 0.51km²，社区现有人口 4610 人、1881 户。驻社区单位有长沙电业局、长沙星电物业管理有限公司、长沙市曙光小学、长沙市用电设备厂等。社区居民从人文环境上看，居民的文化层次、道德素质普遍较高，对政府号召有着较高的互动。

2. 面积要求

选取设计参考区域（20 hm² 以内）展开城市设计。

3. 设计要求

（1）认真收集现状基础资料和相关背景资料，分析城市上一层次规划对基地提出的规划要求，以及基地现状与周边环境的关系，并提出相应的设计主题或设计概念，选择合适的地段进行规划设计，理解具体的设计问题。

（2）提出此次城市设计的整体目标和意图，确定建筑容量，确定城市设计的基本要素。提供具体的功能布局方案，并结合街区外部空间组织、天际线控制、景观开放空间等城市设计框架，包括建筑布局、绿地水系系统、交通系统组织和地下空间利用方案等。

（3）分析并优化本规划区内部居民的交通出行方式，布局道路交通系统，结合其他要素综合考虑道路景观的效果，必要时设计相应的道路断面效果图，确定停车场的类型、规模和布局。

（4）分析并确定本街区公共建筑的内容、规模和布局方式，表达其平面组合体型和室外空间场地的设计构思，公共建筑的配置应结合当地居民生活水平和文化生活特征，结合原有公建设施一并考虑。

（5）绿化系统规划应层次分明、概念明确，与街区功能和户外活动场地统筹考虑，必要时应提出相应的环境设计图，绿化种植设计应结合本土特色。

（6）鼓励同学在对基地现状进行全面分析的基础上，结合本地区的自然条件、生活习惯、历史文脉、技术条件、城市景观等方面进行规划构思，提出优美舒适、有创造性的设计方案。

（7）图纸表达规范：图纸能充分表达设计的理念和创意，且需包含设计文字性说明（包括具体的指标核算说明）。

课题三

长沙火车站周边区域城市设计
组织人：许昊皓

Topic 3: Urban Design of the Area around Changsha Railway Station

一、教学目的

基于城市特定区域的城市历史文化经纬，通过城市设计的理念和技术手段来分析、构想、调整、改造并优化多样类型的城市空间。课题鼓励同学敞开并创新城市特定地段改造保护的思路，认真调研，同时汲取历史先例中城市设计所采用过的优秀技术方法和经验。课程希望同学通过学习，初步掌握基本的城市设计原理，并具有一定的城市问题的解决能力，其中包括现场调研 — 分析问题 — 提出问题 — 设计回应（形态、功能、历史、文化、景观、交通等方面）等。

通过 10 周的设计课程，完整体验学习城市设计的方法和工作流程，理解城市设计的理性实践过程；掌握城市设计方案表达的方法和技能，了解城市设计成果编制的一般过程、图纸要求和形式。

二、基地环境

规划设计区域以长沙火车站为核心，东至东二环路，西至车站中路，北至远大一路，南至人民中路，另包括阿波罗商业广场，规划研究面积约为 120hm²。

每个小组都要根据调研自行选择约 10~20hm² 的街区，进行城市设计方案设计（含 1：500 总平面设计）。

三、设计任务

1. 初步了解城市重要开放空间节点的设计原理与方法，通过研究地段之城市空间历史发展变化脉络及其价值，对本项目基地内空间问题提出解决方法或发展策略。
2. 客观评估"火车站"大型交通建筑的大体量及现代交通建筑对城市空间造成的空间割裂、交通绕行及市民活动公园不足的矛盾现实，提出火车站周边发展的整体谋划。
3. 长沙火车站周边的活力引导和居民生活合理组织：长沙火车站是长沙人生活工作的重要记忆场所。通过对场地现有居民与外来人员的需求分析及生活现状、业态的调研，优化整理现状建筑，进行初步的建筑策划，对地段建设项目提出具有可行性的安排，并提出城市设计建议。

4. 以相关上位规划条件为依据，探讨该地段改造利用的潜力、可能性与方式、方法，寻找符合历史文化特征的高效率开发模式，提出调整、优化和改善场地城市空间环境的城市设计方案。

四、成果要求

1. 城市公共空间调研分析报告文本

形式：全班合作大模型 1 个，A4 文本若干页，小组课堂汇报 1~2 次。

运用城市形态调查分析方法，分析选定的城市空间，描述其形态特征、要素构成、价值、问题等，分析其成因及提出设计对策。

成果要求：

（1）基础现状条件分析图。
（2）场地实体模型 1：1000 。
（3）场地 SKETCHUP 模型。
（4）整体空间结构图解。
（5）专题特色研究分析图。

2. 城市设计方案

小组成果，包括地段尺度和建筑尺度。

形式：A1 展板 4 张，A3 汇报文本 1 份，课堂汇报讨论若干次。

图纸内容可包含区位分析图、用地现状图、总平面图、鸟瞰图、城市形态历史演进分析图（场地）、空间结构、保护策略和改造分析图、用地功能分析图、体量高度分析图、步行网络分析、动静态交通分析、开放空间系统分析、重点节点大比例尺表达、设计说明等。

小组成果要求：

（1）设计说明。
（2）小组设计概念图解。
（3）小组设计成果分析图。
（4）总平面图 1：500。
（5）阶段性工作 SKETCHUP 模型。
（6）主要经济技术指标。

课题四

长株潭融城区城市设计
Topic 4: Urban Design of Changzhutan Region

组织人：彭科

一、设计主题

本次课程以"城市拓展区商业中心城市设计"为主题，旨在考查学生在交通枢纽区塑造城市公共空间品质的能力，鼓励学生以城市设计师的身份探索城市公共空间需求的本质，系统理解城市公共空间的组成与功能，以设计思维、技术手段去解决现实社会问题，使得基地成为城市拓展区，并成为服务于当地居民的新的商业文化中心，以及长株潭融城发展中外地居民进入长沙的新门户。学生需要掌握城市设计的一般原理与方法，促进学生对城市意象的理解，体会城市公共空间、建筑与人之间的密切关系，探索提炼、运用城乡规划知识和建筑营造理论进行创作的路径，进一步培养学生在大尺度公共空间品质提升方面的创造力。

二、概况简介

基地位于长沙市三干两轨发展区域芙蓉大道两厢，基地中心是两条地铁线路的换乘站。由于急速地推进城市化进程，基地周边虽然基本完成城市化建设，但空间品质不高，片区缺乏商业活力。芙蓉大道的快捷化改造在拓宽道路和提高车速上限的同时，极大地割裂和隔绝了芙蓉路两厢的城市活动。城际轨道在通过基地内的先锋站带来长株潭人流的同时也割裂了基地，奥特莱斯广场前的巨大室外停车场也给活力城市的建设带来重大挑战。基地核心是一块三角形的空地，五年前编制的控规将其规划为绿地和停车场用地，控规未考虑先锋站和地铁换乘站等城市建设的新动向。

三、面积要求

本课题是由果子园路、新开铺路与中意路围合而成的区域（约85hm²）。要求学生以小组为单位在该区域内选择用地面积不小于15hm²、不大于30hm²的区域开展城市设计。

四、设计要求

1. 通过城市设计，训练学生综合运用城市规划专业基础知识，提高其对城市规划设计内涵的全面认识和规划分析、研究与表达的综合能力。

2. 引导学生熟悉、掌握与城市拓展区TOD建设相关的法律、法规、标准和技术规范，充分考虑基地内城轨站、地铁站、停车场之间的交通联系。探索以人为本、步行友好城市的空间对策。

3. 提高协调解决城市用地功能的能力，促进城市资源的高效利用，真正做到针对城市资源的补短板与促发展。

4. 促使学生了解城市开发和建设的基本程序、设计程序、编制各阶段设计文件的目的、任务、要求与深度。

课题五

岳麓山大科城核心区城市设计
Topic 5: Urban Design of the Core Area of Yuelushan Dake City

组织人：蒋甦琦

一、教学要点

教学要点之一是在与建筑空间与建筑设计思维的比较中，建立城市公共空间和城市设计思维的方法；之二是了解城市设计的多元利益主体与多元目标的特性，并能够进行简化的功能策划；之三是通过城市调研、文本阅读等，理解城市空间和城市中的"人"及其生活；之四是掌握城市设计和表达的方法，将设计概念贯彻落实到节点空间。

二、规划区域

在岳麓山大科城核心区规划范围中自选 20~30hm² 的土地作为规划设计范围（给出划定红线的理由）。

解决问题：在综合、整体设计的基础上，聚焦于城市公共空间在某一维度下的问题。

1. 本课题中的城市公共空间类型以街道和广场为主。
2. 城市公共空间维度下的问题包括但不限于：
（1）形态维度：城市类型、城市形态等方面问题。
（2）认知维度：场所记忆、城市意象等方面问题。
（3）社会维度：社会生活（后疫情）、文化等方面问题。
（4）视觉维度：城市肌理、风貌、美学等方面问题。
（5）功能维度：土地功能策划、城市环境生态等问题。
（6）时间维度：城市活动与空间的动态变化等问题。

三、教学要求

对应于四个教学阶段，将教学任务分解为四部分，四个部分的作业详见各阶段具体的作业要求。

1. 调研城市
（1）调研方法：快速人类学调研方法、城市设计的实地调研方法、档案查询方法、空间句法等方法。
（2）调研内容：岳麓山大科城核心区公共空间（街道、广场）分析、分类；通过街道客观环境分析和使用者行为与主观感受，对现状大科城主要城市公共空间问题进行梳理，或者对问题溯源，或者对问题提出干预设想。
（3）文本阅读：要求精读《城市设计的维度》，并写出读书报告。

（4）案例分析：针对与调研中发现的问题相适应的案例进行分析。

2. 构想城市
比较建筑设计与城市设计的思维方式的异同，了解城市公共空间的特质，针对问题展开对于城市的构想。
（1）上位评估：对上位规划进行评价，是否有修正的必要。
（2）多元目标：在评估上位规划的基础上，基于多元主体，提出多元目标。
（3）设计概念：以目标为导向，建立问题—概念—反思的逻辑线，设计概念应对于城市公共空间的某一维度。
（4）功能策划：在问题—概念—反思的基础上，进行土地功能策划，提出策划任务书。（外聘教师）

3. 设计城市
（1）总体设计：在目标、概念、策划的基础上进行总体设计，包括空间结构图、各专项的结构图、总平面图、竖向设计等。
（2）节点设计：在总体设计基础上，以城市空间类型和空间叙事为设计方法，选择 2~3 个重要的城市公共空间节点进行设计。

4. 评估城市
（1）城市空间的设计后评估。
（2）人文层面的评价 —— 场所精神。
（3）人文层面的评价 —— 空间正义。

四、设计相关资料

1. 必读书：《城市设计的维度》。
2. 选读书
（1）[英]彼得·琼斯《交通链路与城市空间 —— 街道规划设计指南》，中国建筑工业出版社
（2）王建国《城市设计》，中国建筑工业出版社
（3）庄宇《城市设计实践教程》，中国建筑工业出版社
（4）[美]阿兰·B.雅各布斯《伟大的街道》，中国建筑工业出版社
3. 相关资料：原有岳麓山大学科技城规划资料。

课题一：凤凰县老城区旧城更新城市设计

Topic 1: Urban Design of Phoenix County Old Town Old Town Updates

● **总平面图**　　　General plan

规划总面积：26.67hm²
建设用地总面积：15.95hm²
山体面积：10.92hm²
建筑总基底面积：1.94万m²
建筑总面积：17.24万m²
绿地总面积：1.67万m²
容积率：1.08
绿地率：25%
建筑密度：38%

I 门户广场
II 民俗博物馆
III 旅游综合服务中心
IV 戏台
V 民俗商业街
VI 滨江驳岸
VII 眺望台
VIII 综合商业区
IX 水景公园
X 社区服务中心

1-1剖面

● **鸟瞰图**　　　Airscape

2-2剖面　　　　　　　立面图

设计题目：凤起一脉·凰落民家
　　　　　—— 基于有机更新理论的凤凰县老城区旧城更新城市设计
指导老师：向辉、姜敏
学　　生：贾清利、钱欣茹

Design topic: Based on the Theory of Organic Renewal, the Old City of Phoenix County Updated the Urban Design

Instructors: Xiang Hui，Jiang Min

Students: Jia Qingli，Qian Xinru

● **前期分析**　　**Prophase analysis**

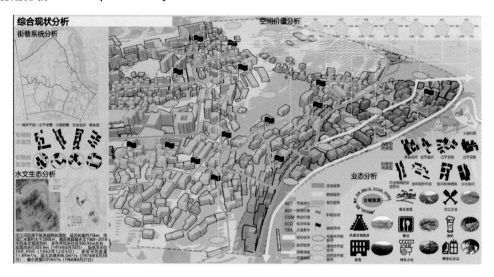

● **设计策略**　　**Design strategy**

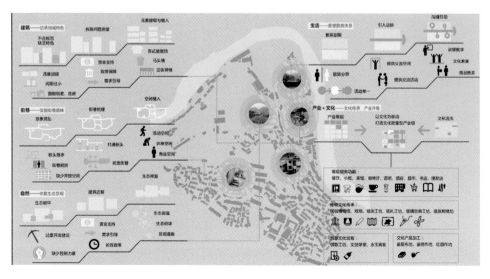

● 总平面图　　General plan

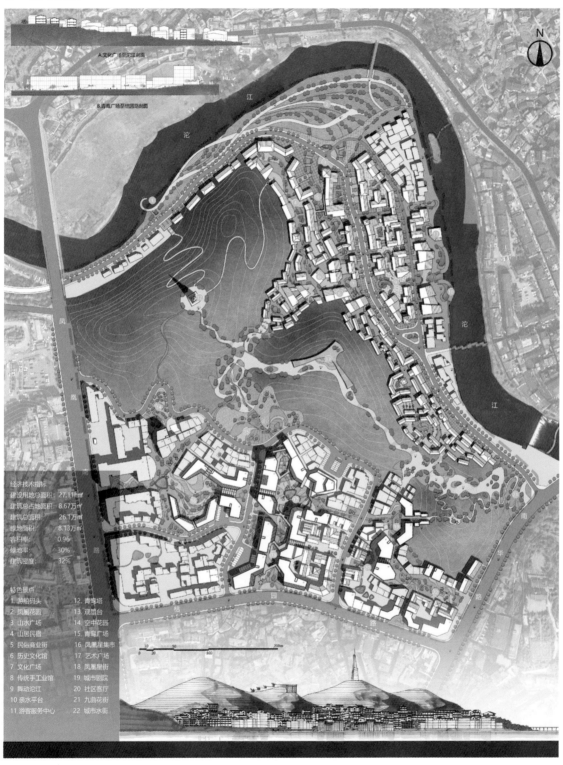

经济技术指标
建设用地总面积：27.11万㎡
建筑总占地面积：8.67万㎡
建筑总面积：26.1万㎡
绿地面积：8.13万㎡
容积率：0.96
绿地率：30%
建筑密度：32%

特色景点
1. 游船码头
2. 凤凰花园
3. 山水广场
4. 山居民宿
5. 民俗商业街
6. 历史文化馆
7. 文化广场
8. 传统手工业馆
9. 舞动沱江
10. 亲水平台
11. 游客服务中心
12. 青窑塔
13. 观景台
14. 空中花园
15. 青窑广场
16. 凤凰星集市
17. 艺术广场
18. 凤凰星街
19. 城市剧院
20. 社区客厅
21. 九曲花街
22. 城市水街

设计题目： 青鸾 —— 基于最大化开放性体验原则的丘陵城市设计
指导老师： 向辉、姜敏
学　　生： 任安之、张达

Design topic: Hilly City Design Based on the Maximizing Openness Experience Design Principle

Instructors: Xiang Hui，Jiang Min

Students: Ren Anzhi，Zhang Da

● **前期分析**　　**Prophase analysis**

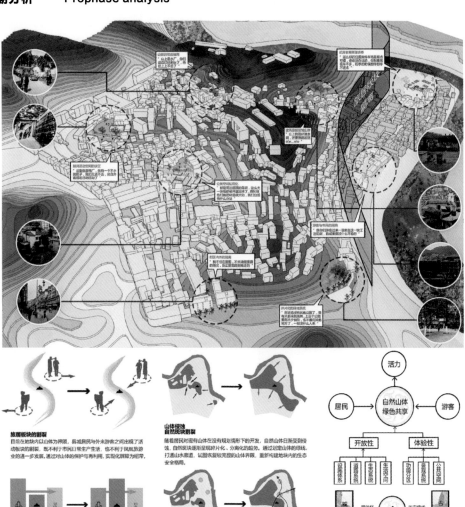

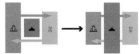

旅居板块的割裂
目前在地块内现有山体周边建筑定案，易城居民与外来游客之间出现了活动板块的割裂，既不利于市民日常生产生活，也不利于凤凰旅游业的进一步发展，通过对山体的保护与再利用，实现化屏障为纽带。

山体的可感知性低
地块内现有山体周边建筑密集，公共空间缺失，道路网密度低，优质景观资源利用率不高，游客和居民都难以在山体周边进行活动，更缺少对自然山体的感知能力。作为丘陵城市，通过设计充分利用自然山体，打造凤凰山-水-城一体的优美画卷。

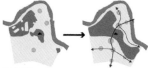

山体侵蚀
自然图块割裂
随着居民对现有山体在没有规划情形下的开发，自然山体日所受到侵蚀，自然图块逐渐呈现碎片化、分离化的趋势，通过划定山体的绿线，打通山水廊道，试图恢复较完整的山体界限，重新构建地块内的生态安全格局。

开放空间缺失
路网密度低
现有场地内部除沱江泊岸之外，少有供市民和游客活动的开放空间，且现有道路难以将开放空间串联。设计在依托山体打造新的开放空间的基础上，提升区域道路网密度，创造良好的开放空间体系。

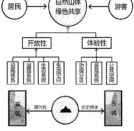

我们提出"青鸾"的概念，以期提升凤凰县内自然山体的开放性与体验性，打破或现在山体所起的屏障隔离作用，化屏障为纽带，提升自然山体在人们视知觉中的地位，将外来游客和本地市民的活动引导向山体，与青山雨林实现绿色共享，由共享而提升山体地区的活力。在古朴的古城风凤凰和宜居的现代风凤凰之上，打造自然生态的绿色凤凰。

● 设计策略　　Design strategy

✔ 策略一：山体的保护与高效利用

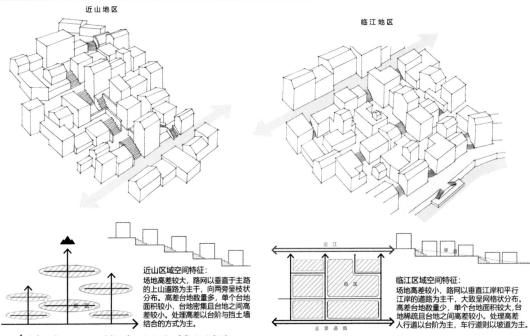

近山地区　　　　　　　　　　　　　　临江地区

近山区域空间特征：
场地高差较大，路网以垂直于主路的上山道路为主干，向两旁呈枝状分布。高差台地数量多，单个台地面积较小，台地密集且台地之间高差较小。处理高差以台阶与挡土墙结合的方式为主。

临江区域空间特征：
场地高差较小，路网以垂直江岸和平行江岸的道路为主干，大致呈网格状分布。高差台地数量少，单个台地面积较大，台地稀疏且台地之间高差较小。处理高差人行道以台阶为主，车行道则以坡道为主。

✔ 策略二：依水而生的凤凰

1.建筑紧邻水岸
江边建筑与水边的距离约为4到5米，优点是建筑的亲水性较强，缺点是空间缺少变化和趣味性，不适于过多使用。由于能够高效利用土地，故在凤凰古城后期加建的区域比较常见。

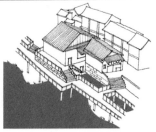

2. 建筑+廊道
采用建筑与廊道复合的空间形式，既增加了建筑的亲水性，又使得空间富有趣味性。在古城核心的地段使用，可识别度高，是游客游览体验的重要节点。

3.江边游园
江边游园呈带状分布，宽度约为6到10米，在古城内增加了自然的元素，同时也起到疏散江边游客的作用，在雨季时，部分的游园会被淹没，但并不影响游客的安全，在建筑与江边距离较远时可以考虑设置滨江游园。

4.吊脚楼
吊脚楼景观处在凤凰古城最为核心的位置，是凤凰古城的标志性景观，与后期加建的建筑相比，能够更好得适应沱江水位的变化。

5.江边标志物
在凤凰古城保护较好的核心地段，有较多的标志物景观，如白塔、北城门楼、老城墙等，这些标志物均具有较高的可识别性，在新建区域值得借鉴。

6.多层廊道
在江边高差变化较大的区域，采用多层廊道的手法进行处理，可以良好地消解高差，避免过多集中的台阶，也提升的空间的趣味性。

鸟瞰图及节点效果图　Aerial view and effect drawings of nodes

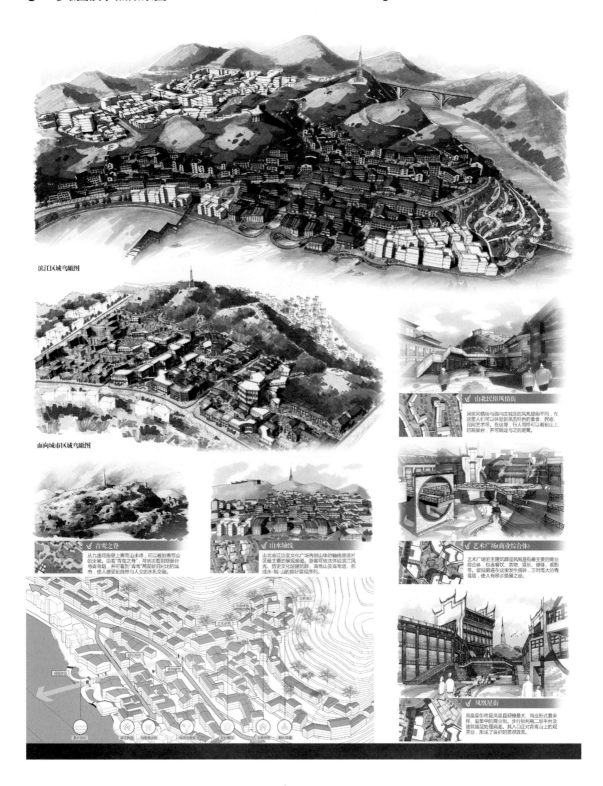

滨江区域鸟瞰图

面向城市区域鸟瞰图

青鸾之脊

从九连花街望上青鸾山未峰，可以看到青鸾山的全景。俯瞰"青鸾之脊"，可依次看到观景台与鸾高塔，并可看到"青鸾"两翼新旧对比的城市，使人感受到自然与人文的水乳交融。

山水轴线

山北由江边至文化广场再到山体的轴线是该片区最主要的景观廊道。游客可依次体验滨江风光、历史文化旅游建筑群、青鸾山及青鸾塔，形成水-城-山的良好景观序列。

山北民俗风情街

民俗风情街与面向主城区的凤凰星街不同，在这里人们可以体验到凤凰别色的美食、民造、民间艺术等。在这里，行人同样可以看到山上的观景台，并可确定与之的距离。

艺术广场（商业综合体）

艺术广场的主建筑群是凤凰星街背后最主要的商业综合体，包含餐饮、购物、娱乐、健体、观影等。景观廊道在这里发生偏转，正对庞大的青鸾塔，使人有移步换景之感。

凤凰星街

凤凰星街将最是凤凰县县积最大、商业形式最多样、最集中的商业街。步行街利用二层平台及建筑错层处理高差。其入口正对青鸾山上的观景台，形成了良好的观景效果。

229

● **总平面图**　　General plan

设计题目： 慢缘山行 谋以安居 —— 基于慢行理念下的山地智慧城市设计
指导老师： 孙亮
学　　生： 罗思阳、吴俊沛

Design topic: Mountainous Smart City Design Based on the Concept of Slow Traveling

Instructors: Sun Liang

Students: Luo Siyang，Wu Junpei

● 设计说明

我们希望基于慢行理念，在东瓜山的场地内置入一个安全、便捷、友好的慢行网络，串联起不同类别、不同层级的活动场所，同步解决区域内的交通混乱以及使用者活动场所匮乏的问题。同时，整个系统顺应东瓜山原本地形，尊重原有空间体系，通过较少的改变使这片以老旧社区为主的区域重新焕发活力，更加舒适宜居。

Design notes

Based on the concept of slow travel, we want to build a safe, convenient and friendly slow-track network in Dongguashan to connect different types of different levels of activity venues, simultaneously solve the problem of traffic chaos in the region and the lack of user activity space. At the same time, the whole system conforms to the original terrain of Dongguashan, respects the original space system, and through fewer changes, makes this area, which is dominated by old communities, revitalized and more comfortable and livable.

● 前期分析　　Prophase analysis

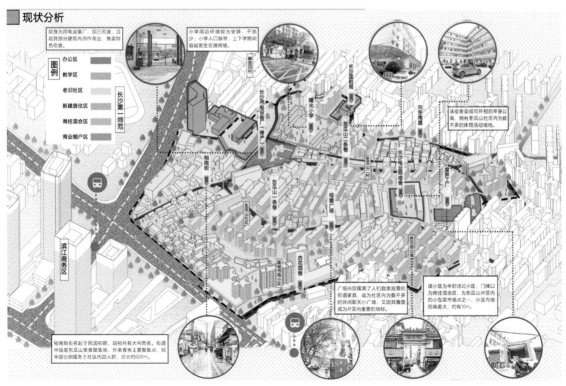

设计内容　Design content

慢行系统

- ▬▬▬▬▬▬▬ 主轴 综合
- ▬▬▬▬▬ 次轴 活动/通行
- ▬▬▬▬▬ 次轴 功能/通行

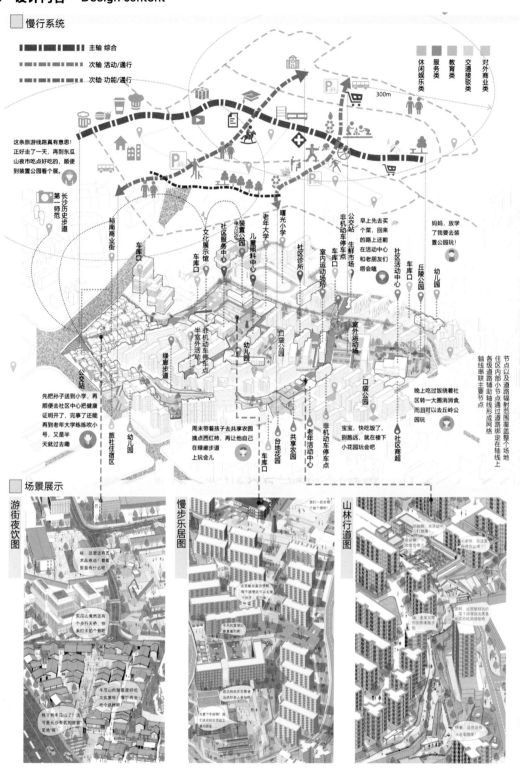

休闲娱乐类　服务类　教育类　交通接驳类　对外商业类

300m

这条旅游线路真有意思！
正好走了一天，再到东瓜
山夜市吃好吃的，顺便
到装置公园看个展。

长沙历史步道
第一师范

节点以及道路辐射范围覆盖整个场地
住区内部小节点通过道路织定在轴线上
各级道路辅助轴线形成网络
轴线串联主要节点

裕南商业街
车库口
文化展示馆
社区服务中心
装置公园
老年大学
曙光小学
儿童照料中心
非机动车停车点
公交站
生鲜市场
车库口
社区诊所
室内运动场所
室外运动场
社区活动中心
车库口
丘陵公园
幼儿园

妈妈，放学了我要去装置公园玩！

早上先去买个菜，回来的路上还能在活动中心和老朋友们唠会嗑

非机动车停车点
半室外活动
绿廊步道
幼儿园
口袋公园
口袋公园
室外运动场
社区商超

晚上吃过饭绕着社区转一大圈消食，而且可以去丘岭公园玩

公交站

先把孙子送到小学，再顺便去社区中心把健康证明开了，完事了还能再到老年大学练练吹小号，又是半天就过去咧

旅社住宿区
幼儿园
周末带着孩子去共享农园摘点西红柿，再让他自己在绿廊步道上玩会儿
台地花园
共享农园
车库口
老年活动中心
非机动车停车点

宝宝，快吃饭了，别跑远，小花园玩会吧

场景展示

游街夜饮图

慢步乐居图

山林行道图

● 鸟瞰图　Aerial view

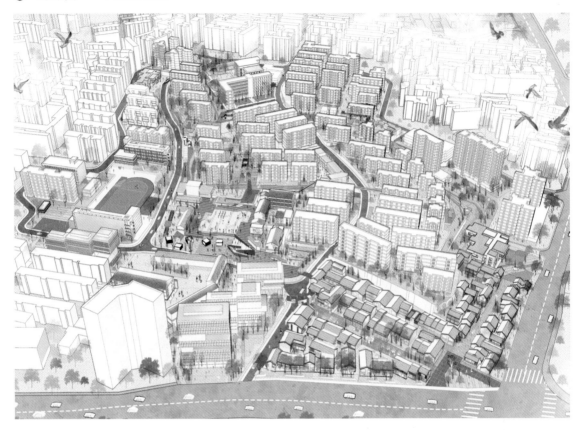

● 节点分析图　Node analysis

■ 节点平面图 1:500　　　　　　　　　　■ 节点剖面图 1:500

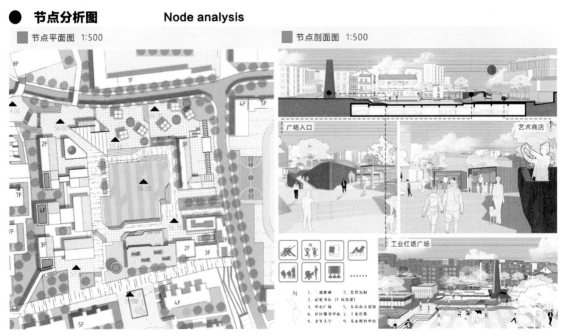

东瓜山培养基 03

指导教师 陈焯
组员 李杼欣 高原

经济技术指标
规划用地面积：26.29hm²
总建筑面积：320494 m²
建筑密度：29.7%
容积率：1.22
绿地率：37%
地面停车位：50个
平均层数：5.8层

1. 宅前示范合作绿地
2. 高低公园
3. 街角广场
4. 社区绿色活动中心
5. 楼间绿廊

6. 景观谷
7. 主题种植区
8. 休闲露台
9. 架空廊道

10. 生态种植产业园
11. 休闲商业点
12. 慢行步道

13. 社区活动中心
14. 生活绿地广场

15. 社区共享温室
16. 种植培育中心
17. 东瓜山大院

总平面图 1:1500

设计题目: 东瓜山培养基 —— 基于生态修补与都市种植语境下的东瓜山片区城市设计

指导老师: 陈煊

学　　生: 高原、李杼欣

4

Design topic: Dongguashan District Urban Design in the Context of Ecological Restoration and Urban Planting

Instructors: Chen Xuan

Students: Gao Yuan，Li Zhuxin

● **前期分析**　Prophase analysis

● **生态策略**　Ecological strategy

235

● 设计策略　Design strategy

■城市绿地系统 URBAN GREEN SPACE SYSTEM

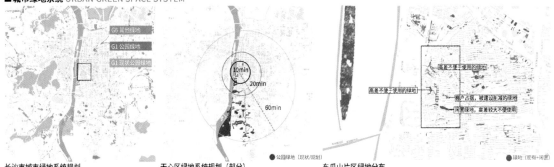

长沙市城市绿地系统规划

城市绿地系统建设中，以自然资源的系统性建设为主，因此河西的绿地面积要大于河东，以G6属性绿地最明显，场地周围很少大型绿地

天心区绿地系统规划（部分）

作为高密度居住区，天心区的公园绿地，现状绿地，其他绿地三部分绿地中位于东瓜山短时生活圈内的占很少部分

东瓜山片区绿地分布

东瓜山的绿化覆盖率（卫星图）只有14%，对于一个高密度居住片区来讲是远远不能达标的，加上城市绿地系统的建设模式与居住区的关联是这远不够善至缺失的

■设计逻辑 DESIGN LOGIC

大节点串联，向城市开放　　　　绿地修补，向外扩张　　　　小节点改造，联合打造

■总平面结构 GENERAL LAYOUT STRUCTURE

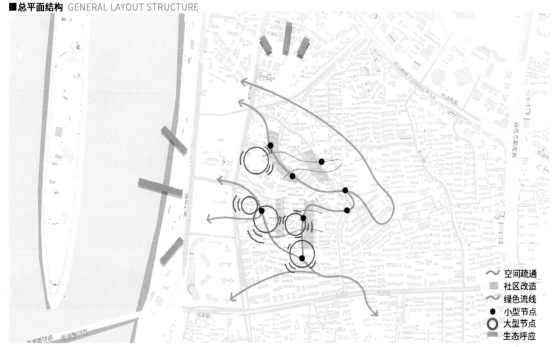

〜 空间疏通
■ 社区改造
〜 绿色流线
● 小型节点
○ 大型节点
■ 生态呼应

● 总平面分析　　Analysis of general plan

▉	宅间改造居民楼
▉	功能绿色游园
▉	高差改造居民楼
▉	宅间改造居民楼
▉	社区活动中心
▉	商业
▉	幼儿园

建筑/地块功能分析

● 片区入口

交通流线分析

● 节点设计　　Nodes design

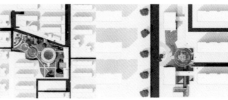

东瓜山种植中心
种植管理
种植研究
社区对接
产业合作中心
绿色社区协会

实验种植环
游园种植环
种植样本示范
穿行观光
社区后花园
开放广场
城市峡谷游园

公共广场改造
坡地步道
阶梯/平台

社区活动小广场
绿色空中步行环
绿色民宿体验
大院观景平台
东瓜大院集场
大院空中市区
大院景观展览中庭

大院主题种植核
大院管理
主题种植范本展览
城市活动对接
大院管理中心
商业植入
东瓜山地标特色建设
大院推广模式

大院集
东瓜山院下集市
东瓜产业园
大院儿童乐园
东瓜大院节

大院广场
东瓜步行广场
东瓜公园
东瓜山社区剧场
东瓜山城市园口

东瓜山社区支持种植基地/Planting Base

原用地南面的闲置条状绿地，是南北社区间的巨大隔断，利用其天然的坡地改造为东瓜山社区种植总基地，连接西边的绿色峡谷直通城市空间

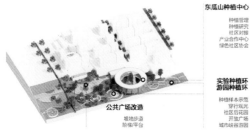

东瓜山大院/Biiiig Courtyard

东瓜山片区中部的大型"盆地"式地块，被多个社区以及商业街围绕，交通核心。结合巨大的高差进行综合体建设，通过垂直空间划分衍生出民宿庄园，居民广场，种植农园等一体的体量

● 总平面图　　General plan

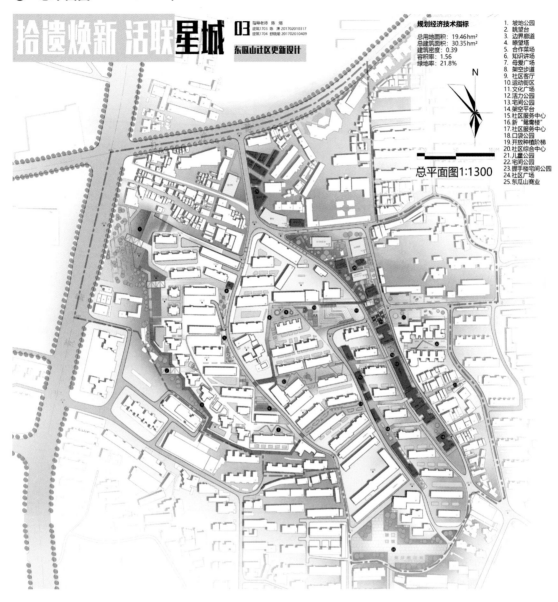

拾遗焕新 活联星城 03

东瓜山社区更新设计

指导老师 陈 焰
建筑1703 陈 澳 201702010317
建筑1704 舒靖旭 201702010409

规划经济技术指标

总用地面积：19.46hm²
总建筑面积：30.35hm²
建筑密度：0.39
容积率：1.56
绿地率：21.8%

N

总平面图1:1300

1. 坡地公园
2. 眺望台
3. 边界廊道
4. 瞭望塔
5. 合作菜场
6. 知识讲场
7. 母爱广场
8. 架空步道
9. 社区客厅
10. 运动街区
11. 文化广场
12. 活力公园
13. 宅间公园
14. 架空平台
15. 社区服务中心
16. 新 "鸳鸯楼"
17. 社区服务中心
18. 口袋公园
19. 开放种植阶梯
20. 社区综合中心
21. 儿童公园
22. 宅间公园
23. 握手楼宅间公园
24. 社区广场
25. 东瓜山商业

● 规划分析　　Planning analysis

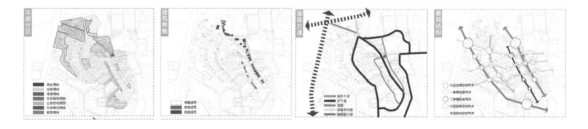

设计题目： 拾遗焕新 活联星城 —— 东瓜山社区更新设计
指导老师： 陈煊
学　　生： 陈潇、舒晓瑜

Design topic: Dongguashan Community Update Design

Instructors: Chen Xuan

Students: Chen Xiao，Shu Xiaoyu

● **前期分析**　Prophase analysis

问题挖掘

从观察场地现象入手，总结过后，我们从五个角度记录了场地内的种种现象，并进行还原，提取出场地真实存在的问题。其中，"商业街区"包含的现象有商业街交通混乱、棚户区脏乱差、巨大高差带来的边界空间的失效、商业闹市；"公共空间"包含的现象有公共空间私有化、居民自主营造、边角空间利用率低、规划公共空间使用率低；"社区工作"包含的现象有居委会的管理不够、社区活动室的冷落；"环境特征"包含的现象有高差复杂、社区围墙多、社区景观被破坏；"日常活动"所包含的现象有居民交往方式的单一（麻将）、儿童游乐空间的缺失等。在这些现象的背后，我们挖掘出了真实存在的问题，具体问题包括社区与城市、社区内部两大问题；细分"社区与城市"包含商住矛盾和物理边界的问题，"社区内部"包含公共空间和社区共建的问题

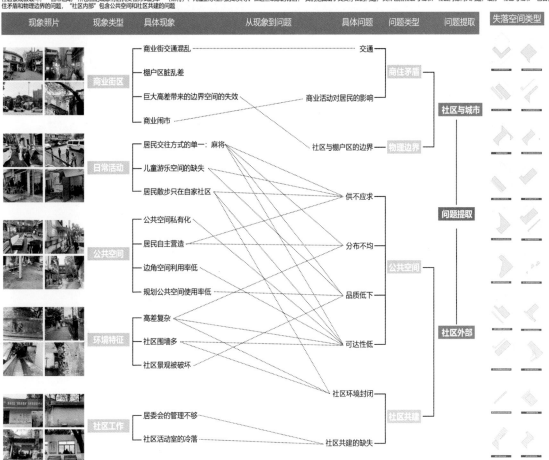

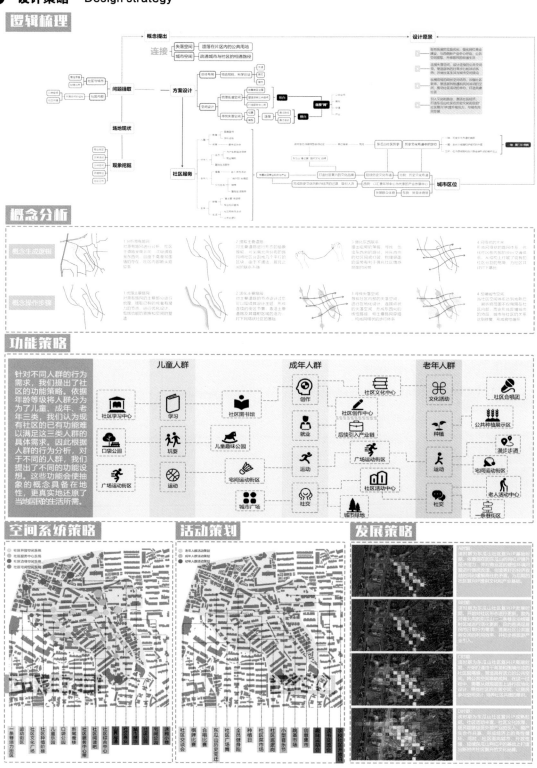

设计策略　Design strategy

● 鸟瞰图　　Aerial view

拾遗焕新 活联**星城** 04
东风山社区更新设计

● 节点分析及效果图　　Nodes analysis and design sketch

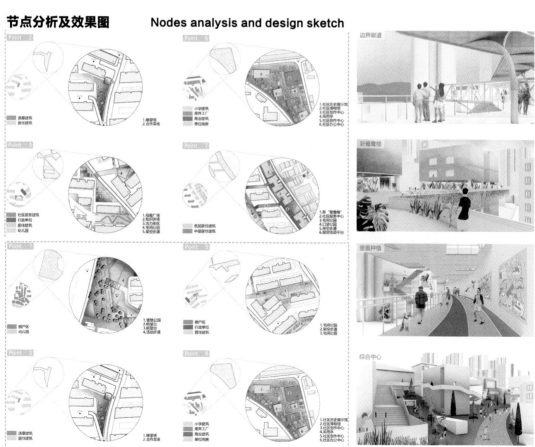

边界廊道

新建筑楼

垂直种植

综合中心

课题三：长沙火车站周边区域城市设计
Topic 3: Urban Design of the Area around Changsha Railway Station

● **总平面**　　　General plan

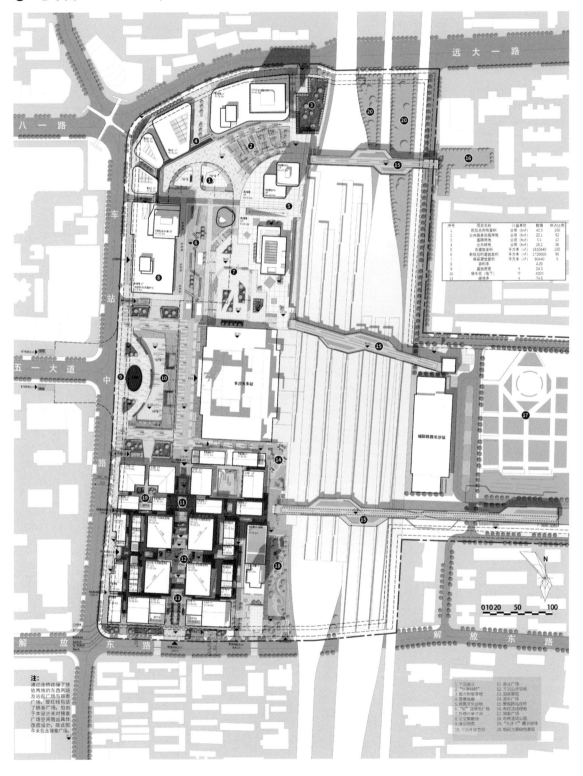

6

设计题目： 双核站城 都市引擎 —— 城市更新开发视角下的长沙火车站片区城市设计
指导老师： 许昊皓、彭科
学　　生： 赵英伦、王佳奇

Design topic: Changsha Railway Station Area Urban Design from the Perspective of Urban Renewal and Development

Instructors: Xu Haohao，Peng Ke

Students: Zhao Yinglun，Wang Jiaqi

● **前期分析**　Prophase analysis

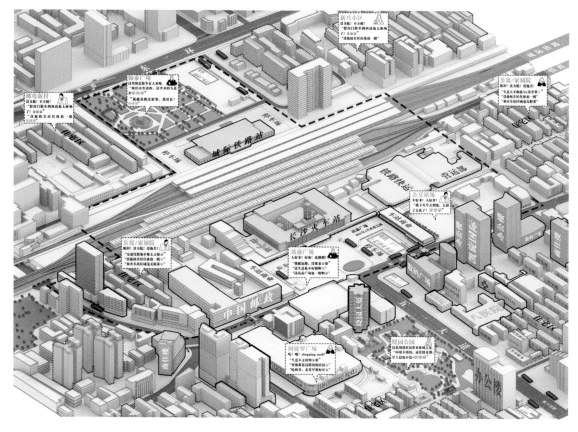

设计策略　　　　Design strategy

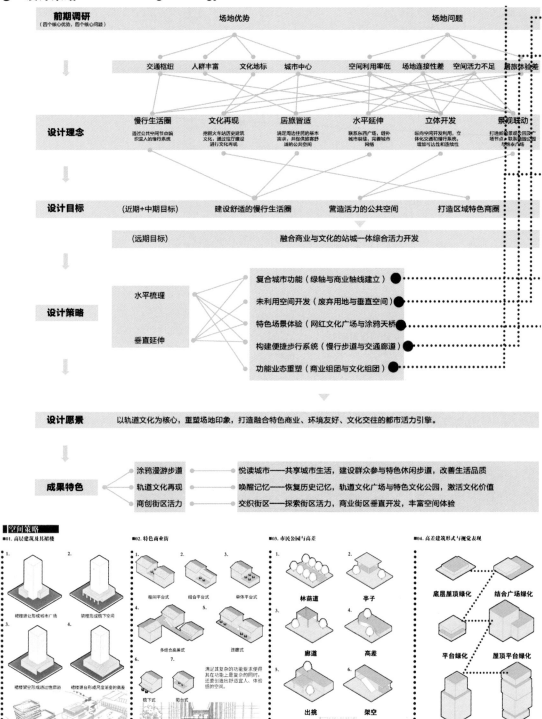

前期调研
（四个核心优势，四个核心问题）

场地优势　　　　　　　　　　**场地问题**

交通枢纽　人群丰富　文化地标　城市中心　　空间利用率低　场地连接性差　空间活力不足　居旅体验差

设计理念

慢行生活圈	文化再现	居旅皆适	水平延伸	立体开发	景观联动
通过公共节点组织置人的慢行系统	挖掘火车站历史建筑文化，通过场厅建设进行文化再现	满足周边住民的基本需求，并提供旅客舒适的公共空间	联系东西广场，缝补城市裂缝，完善城市网络	纵向空间开发利用，立体化交通和慢行系统，增加可达性和连续性	打造新的景观多节点广场，联系周围公园节点等场地

设计目标

（近期+中期目标）　　建设舒适的慢行生活圈　　营造活力的公共空间　　打造区域特色商圈

（远期目标）　　　　融合商业与文化的站城一体综合活力开发

设计策略

水平梳理

垂直延伸

- 复合城市功能（绿轴与商业轴线建立）●
- 未利用空间开发（废弃用地与垂直空间）●
- 特色场景体验（网红文化广场与涂鸦天桥）●
- 构建便捷步行系统（慢行步道与交通廊道）●
- 功能业态重塑（商业组团与文化组团）●

设计愿景

以轨道文化为核心，重塑场地印象，打造融合特色商业、环境友好、文化交往的都市活力引擎。

成果特色

涂鸦漫游步道　　悦读城市——共享城市生活，建设群众参与特色休闲步道，改善生活品质
轨道文化再现　　唤醒记忆——恢复历史记忆，轨道文化广场与特色文化公园，激活文化价值
商创街区活力　　交织街区——探索街区活力，商业街区垂直开发，丰富空间体验

空间策略

01. 高层建筑及其裙楼
1. 2.
裙楼退让形成城市广场
骑楼形成檐下空间
3. 4.
裙楼架空形成通过性街道
裙楼退台形成尺度亲善的商业

02. 特色商业街
相间平台式　组合平台式　单体平台式
多组合底差式　连廊式
桥下式　阳台式
满足其复杂的功能要求使得其在功能上是最复杂的同时，还要创造出舒适宜人、体验感的空间。

03. 市民公园与高差
林荫道　亭子
廊道　高差
出挑　架空

04. 高差建筑形式与视觉表现
底层屋顶绿化　结合广场绿化
平台绿化　屋顶平台绿化
环绕式屋顶平台绿化　错落多层绿化

244

● 效果图　　Design sketch

鸟瞰效果图

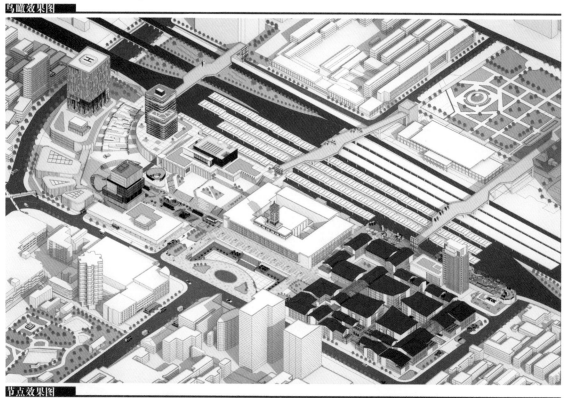

节点效果图

立面图

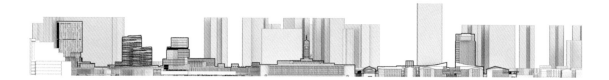

规划策略 Planning strategy

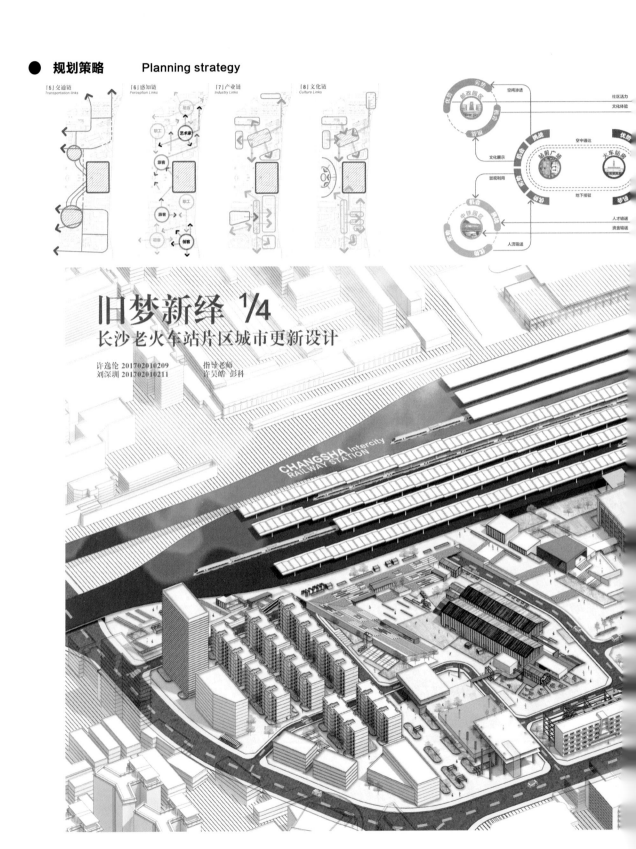

[5] 交通链
Transportation links

[6] 感知链
Perception Links

[7] 产业链
Industry Links

[8] 文化链
Culture Links

旧梦新绎 ¼

长沙老火车站片区城市更新设计

许逸伦 201702010209　　指导老师
刘深圳 201702010211　　许昊皓 彭科

CHANGSHA Intercity
RAILWAY STATION

设计题目：旧梦新绎 —— 长沙火车站片区城市更新设计
指导老师：许昊皓、彭科
学　　生：许逸伦、刘深圳

7

Design topic: Changsha Railway Station Area Urban Renewal Design

Instructors: Xu Haohao，Peng Ke

Students:Xu Yilun，Liu Shenzhen

● 鸟瞰图　　Aerial view

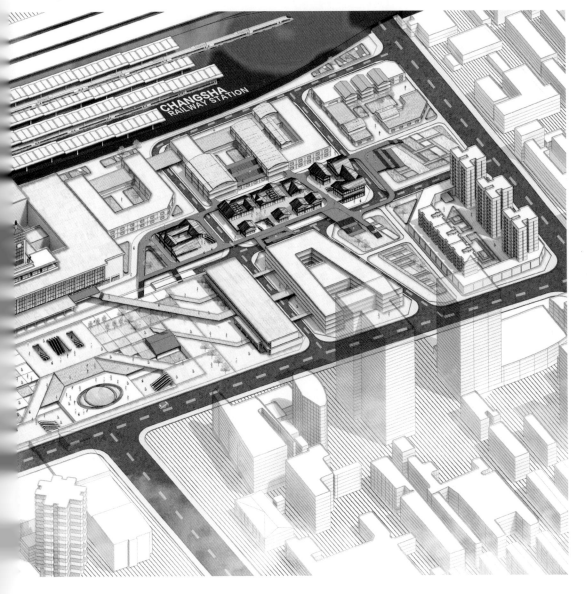

总平面图　　General plan

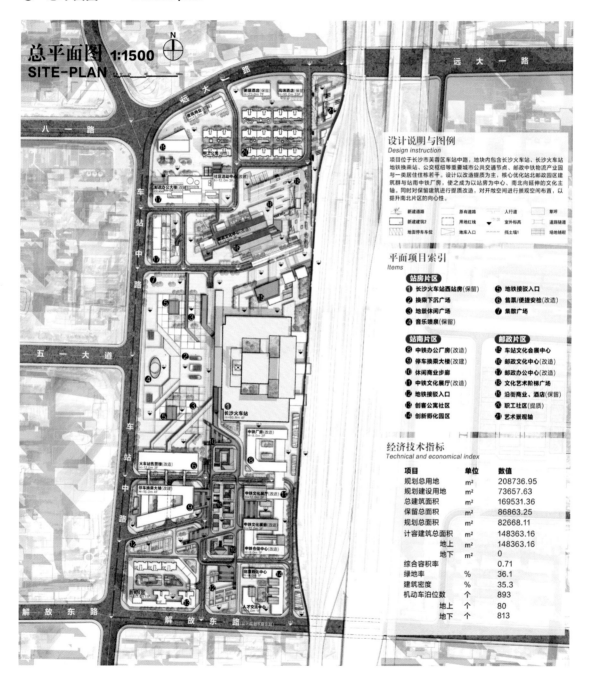

总平面图 1:1500
SITE-PLAN

远大一路

八一路

远大一路

车站中路

五一大道

车站中路

解放东路

设计说明与图例
Design instruction
项目位于长沙市芙蓉区车站中路，地块内包含长沙火车站、长沙火车站地铁换乘站、公交枢纽等重要城市公共交通节点、邮政中铁物流产业园与一类居住住栋若干。设计以改造提质为主，核心优化站北邮政园区建筑群与站南中铁厂房，使之成为以站房为中心、南北向延伸的文化主轴，同时对保留建筑进行提质改造，对开敞空间进行景观空间布置，以提升南北片区的向心性。

新建道路	原有道路	人行道
新建建筑2	用地红线	道路铺装
地面停车车位	地库入口	场地铺砌

草坪
室外标高
挡土墙1

平面项目索引
Items

站房片区
❶ 长沙火车站西站房(保留)
❷ 换乘下沉广场
❸ 地景休闲广场
❹ 音乐喷泉(保留)
❺ 地铁接驳入口
❻ 售票/便捷安检(改造)
❼ 集散广场

站南片区
❽ 中铁办公用房(改造)
❾ 停车换乘大楼(改建)
❿ 休闲商业步廊
⓫ 中铁文化展厅(改造)
⓬ 地铁接驳入口
⓭ 创客公寓社区
⓮ 创新孵化园区

邮政片区
⓯ 车站文化会展中心
⓰ 邮政文化中心(改造)
⓱ 邮政办公中心(改造)
⓲ 文化艺术阶梯广场
⓳ 沿街商业、酒店(保留)
⓴ 职工社区(提质)
㉑ 艺术景观轴

经济技术指标
Technical and economical index

项目	单位	数值
规划总用地	m²	208736.95
规划建设用地	m²	73657.63
总建筑面积	m²	169531.36
保留总面积	m²	86863.25
规划总面积	m²	82668.11
计容建筑总面积	m²	148363.16
地上	m²	148363.16
地下	m²	0
综合容积率		0.71
绿地率	%	36.1
建筑密度	%	35.3
机动车泊位数	个	893
地上	个	80
地下	个	813

● 节点设计　Nodes design

北边艺术园区入口

北边大台阶

北边架空廊道

北边架空廊道

喷泉广场

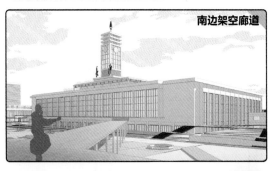
南边架空廊道

步行街入口

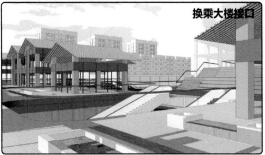
换乘大楼接口

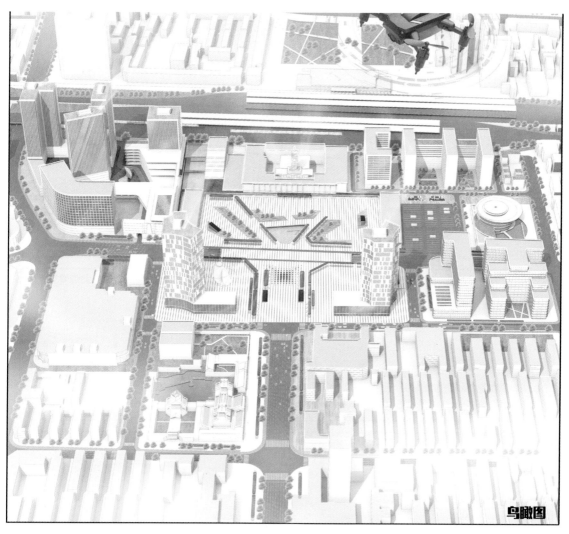

鸟瞰图

18:00PM

19:00PM

21:00PM

I LOVE CS.

8:00AM

设计题目： 城市荧幕 —— 基于地理媒介技术的公共空间重塑
指导老师： 彭科
学　　生： 涂欢乐、杨明达

Design topic: Public Space Remodeling Based on Geographic Media Technology

Instructors: Peng Ke

Students: Tu Huanle、Yang Mingda

● **前期分析**　　Prophase analysis

场地分析 —— 交通接驳分析 —— 接驳混乱，交通堵塞

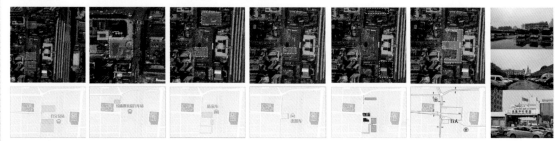

停车分布 —— 停车位缺乏，占用空间　　场地分析 —— 城市形象 —— 形象欠缺，火车站形象淡化

五一大道北立面较单一，缺少城市特色，与南侧立面缺少整体感

空间环境 —— 公共空间浪费

场地内存在厂房、破旧房子等老旧建筑，公共服务设施、景观设施也设置不合理或者老化，大量公共空地闲置或被改成露天停车场，站前广场被出租车、私家车、公交车停满，空间环境亟待提升

南侧立面比较具有特色，建筑立面对塑造城市形象具有帮助意义

火车站建筑形象突出，但广场已经被车辆占据，广场流线混乱

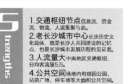

Strengths
1.交通枢纽节点 信息流，资金流，物流，人流重聚于此。
2.老长沙城市中心 长沙历史文化载体，既是长沙人共同建设的记忆点，也是长沙城市发展历程的见证者。
3.人流量大 中南地区交通枢纽，日均客流量5万。
4.公共空间 场地内有晓园公园、站前广场、停车场等大面积公共空间。

Weaknesses
1.低开发强度 土地利用效率低，各类地块内建筑密度大，而土地使用强度却较低，片区范围内建筑以多层为主，也有少量小高和低层。
2.产业低端 各类产业多规划无序生长，缺少区域发展驱动力。
3.交通层级不明确 出租车、地铁、公交等慢速交通、步行、自行车等交通在道路网络分配上未能得到有效梳理。
4.基础服务设施不堪重负，建筑老旧，城市风貌有待提振。

Opportunities
1.城市发展轴 长沙火车站片区处于湘江综合服务轴北部综合发展带的交汇辐射范围，与长沙五一大道连接，形成带机场连线的北部城市发展轴（2014年长沙总体规划）。
2.城市发展极核 长沙火车站片区未来将成为长株潭城市群发展极核（2014年长沙总体规划）。
3.网红城市 长沙定位为媒体艺术之都，打造网红经济。

Threats
1.大量拆建 TAD改成TOD的模式改变，势必涉及大量拆除重建，政府工作难度大。
2.历史与开发 并非白纸的发展背景下，文化传承与全球化建城模式如何融合。
3.城市节奏 难以兼顾慢生活，TOD开发模式主要服务于城市快节奏生活，但也应该满足"慢生活"的需求。

● **交通及功能规划**　Transportation and functional planning

铁路 railway

火车站月台　2F

火车站、广场、人行　1F

主要道路　1F
出租车接驳点、快速通过道路、商业街　-1F

社会车辆停车场　-2F

地铁2号线　-2F

地铁3号线　-3F

道路改造　　　接驳方式

建筑功能

1.人才公寓　建筑面积130436㎡
2.银行公共服务建筑，建筑面积7636㎡
3.历史博物馆公共服务建筑，建筑面积23242㎡
4.旅馆公共服务建筑，建筑面积20000㎡
5.商业建筑，建筑面积118300㎡
6.商务建筑，建筑面积304710㎡
7.火车站交通建筑，建筑面积30000㎡
8.物流上盖商业，建筑面积60000㎡

用地分类

地块1：总面积40000㎡；建筑密度40.8%；容积率5.4
地块2：地块3：总面积21211㎡；建筑密度75.4%；容积率4.1 地块3：总面积50200㎡
地块4：总面积12087㎡；建筑密度44%；容积率6.2
地块5：总面积19380㎡；建筑密度53.8%；容积率7.1
地块6：总面积15350㎡；建筑密度13.6%；容积率0.7
地块7：总面积3950㎡；建筑密度16%；容积率0.7
地块8：总面积31860㎡；地块9：总面积4590㎡；地块10：总面积9900㎡
地块11：总面积2395㎡；地块12：总面积9000㎡；地块13：总面积10600㎡
地块14：总面积6577㎡；地块15：总面积66000㎡；总用地380000㎡

建筑类型

邮政创意园+低层　商业　邮政办公+物流　商业+办公　商业+办公　酒店+低商　博物馆　人才公寓

● 总平面图　General plan

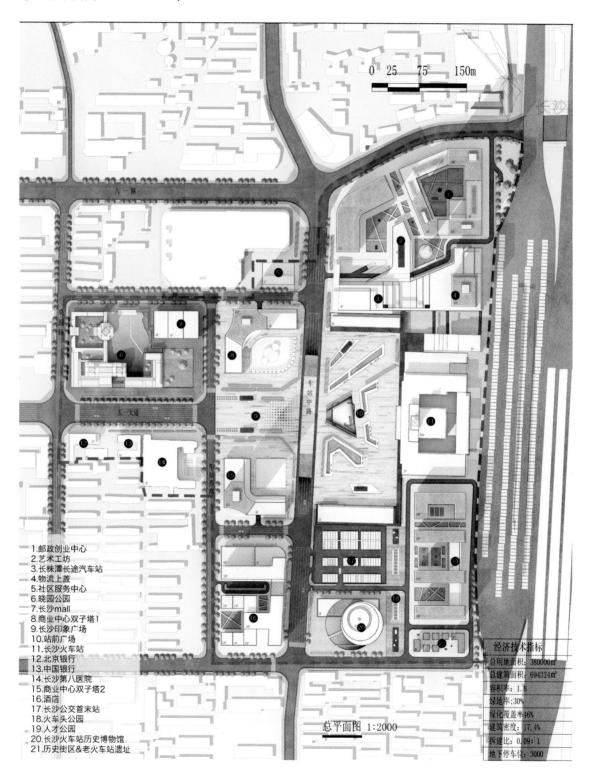

0　25　75　150m

1.邮政创业中心
2.艺术工坊
3.长株潭长途汽车站
4.物流上盖
5.社区服务中心
6.晓园公园
7.长沙mall
8.商业中心双子塔1
9.长沙印象广场
10.站前广场
11.长沙火车站
12.北京银行
13.中国银行
14.长沙第八医院
15.商业中心双子塔2
16.酒店
17.长沙公交首末站
18.火车头公园
19.人才公园
20.长沙火车站历史博物馆
21.历史街区&老火车站遗址

总平面图　1:2000

经济技术指标	
总用地面积:	380000m²
总建筑面积:	694324m²
容积率:	1.8
绿地率:	30%
绿化覆盖率	46%
建筑密度:	17.4%
拆建比:	0.09:1
地下停车位:	3000

课题四：长株潭融城区城市设计
Topic 4: Urban Design of Changzhutan Region

● **总平面图**　　　General plan

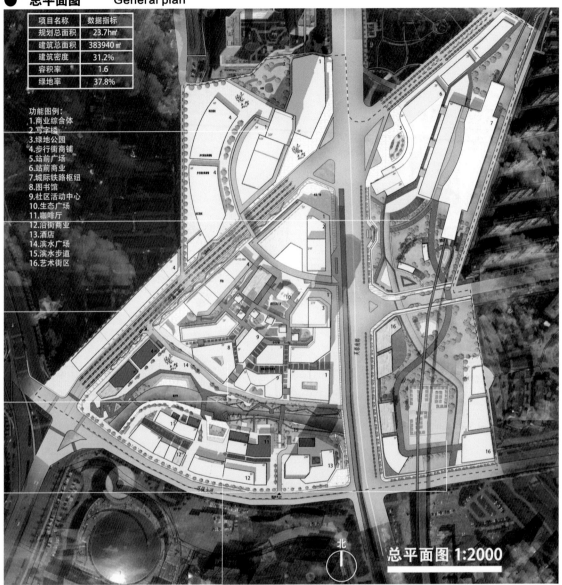

项目名称	数据指标
规划总面积	23.7hm²
建筑总面积	383940 m²
建筑密度	31.2%
容积率	1.6
绿地率	37.8%

功能图例：
1. 商业综合体
2. 写字楼
3. 绿地公园
4. 步行街商铺
5. 站前广场
6. 站前商业
7. 城际铁路枢纽
8. 图书馆
9. 社区活动中心
10. 生态广场
11. 咖啡厅
12. 沿街商业
13. 酒店
14. 滨水广场
15. 滨水步道
16. 艺术街区

北

总平面图 1:2000

● **设计逻辑**　　　Design logic

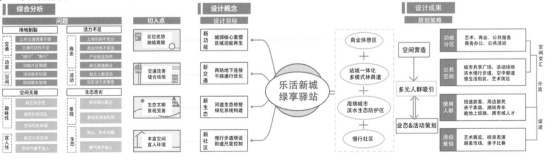

设计题目： 乐活新城·绿享驿站 —— 长株潭融城商业游憩区设计
指导老师： 彭科
学　　生： 廖天怡、王心然、王囡、谢源

9

Design topic: Commercial Recreation Area Design in Changsha-Zhuzhou-Xiangtan Region

Instructors: Peng Ke

Students: Liao Tianyi，Wang Xinran，Wang Nan，Xie Yuan

● **前期分析**　Prophase Analysis

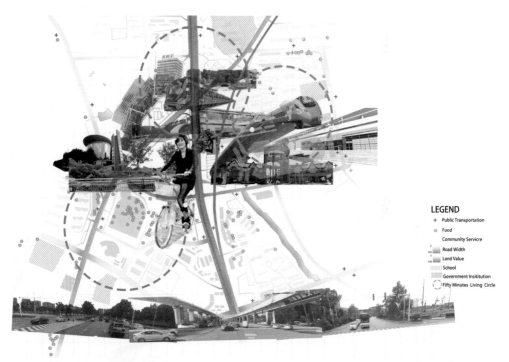

交通分析 Traffic Analysis ——— 交通便利 潜在人群吸引优势

道路车流

研究的基地范围内包括三条主要道路——芙蓉南路、中意一路及环保大道，长沙市区车流和株洲方向车流通过芙蓉南路进入基地。环保大道引入东西方向的车流。场地周边的活动车流可通过次级道路进入基地。

公交

由2018年统计数据看出，长沙市出行人群选择公交的比例仍较大，公交站点较多增加了更多短距离人群到达此处的可能性。

轨道交通

中信广场站为一号线南部重要站点（下一站尚双塘站为二号线终点站），未来地铁五号线（规划中）在此站换乘。据2018年数据显示，地铁1号线日均客运量26.6万人次。一号线途径北辰三角洲、五一广场、黄兴广场、南门口等大人流站点。长株潭城际铁路为长株潭城市群提供快捷、准时、便捷的交通服务，承担3市之间的客流运输。这些乘坐人群将成为此处的潜在使用人群。

人流来向

255

● 设计逻辑　Design logic

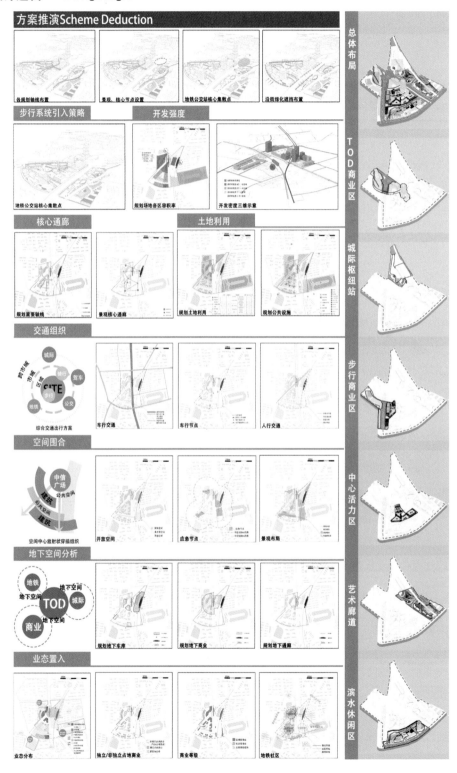

方案推演Scheme Deduction

各规划轴线布置　　景观、核心节点设置　　地铁公交站核心集散点　　沿街绿化遮挡布置

步行系统引入策略　　**开发强度**

地铁公交站核心集散点　　规划场地各区容积率　　开发密度三维示意

核心通廊　　**土地利用**

规划重要轴线　　景观核心通廊　　规划土地利用　　规划公共设施

交通组织

综合交通出行方案　　车行交通　　车行节点　　人行交通

空间围合

空间中心放射状穿插组织　　开放空间　　应急节点　　景观布局

地下空间分析

规划地下车库　　规划地下商业　　规划地下通廊

业态置入

业态分布　　独立/非独立占地商业　　商业等级　　地铁社区

总体布局

TOD商业区

城际枢纽站

步行商业区

中心活力区

艺术廊道

滨水休闲区

● 详细设计　　Detailed design

三角地块详细设计　Midland of Triangle Detailed Design

港子河沿岸详细设计　GangZi River Detailed Design

高架桥下详细设计 Detailed Design Under Viaduct

建筑功能

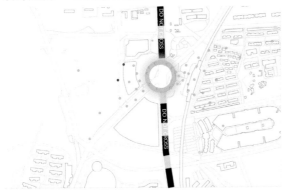

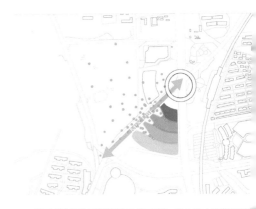

1. 于轨道站口置入地标性天桥，衍生出环形漫步广场，衔接被道路割裂的场地与步行系统。

2. 地景建筑面向西侧居民区"开裂"，结合多功能林荫道，引居民前来消费娱乐。

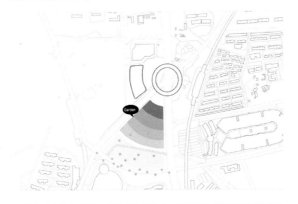

3. 退台式商业综合体由北至南层数递减，以花园地景吸引南侧株潭干道前来游玩的株潭居民。

4. 穿过地景公园的慢行步道衔接环形漫步广场与湿地公园车道建立场地慢行系统。

设计题目：街城之环 —— 南城商业新门户
指导老师：彭科
学　　生：孙凡清、邱子倍、郝洪庆、曾露颖

Design topic: New Commercial Portal in Nancheng of Changsha

Instructors: Peng Ke

Students: Sun Fanqing，Qiu Zibei，Hao Hongqing，Zeng Luying

● **总平面及鸟瞰图**　　General plan and aerial view

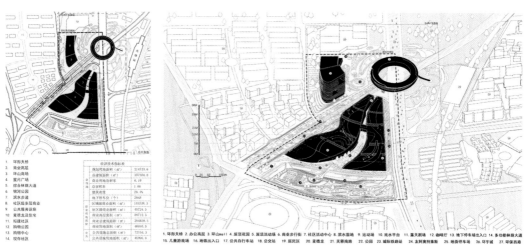

1. 环形天桥 2. 办公高层 3. 环山mall 4. 屋顶花园 5. 屋顶活动场 6. 商业步行街 7. 社区活动中心 8. 滨水湿地 9. 运动场 10. 戏水平台 11. 露天剧场 12. 咖啡厅 13. 地下停车场出入口 14. 多功能林荫大道
15. 儿童游戏场 16. 地铁出入口 17. 公共自行车场 18. 公交站 19. 居民区 20. 麦德龙 21. 芙蓉南路 22. 公园 23. 城际铁路站 24. 友阿奥特莱斯 25. 地面停车场 26. 环宇城 27. 环保西路

多模式林荫大道

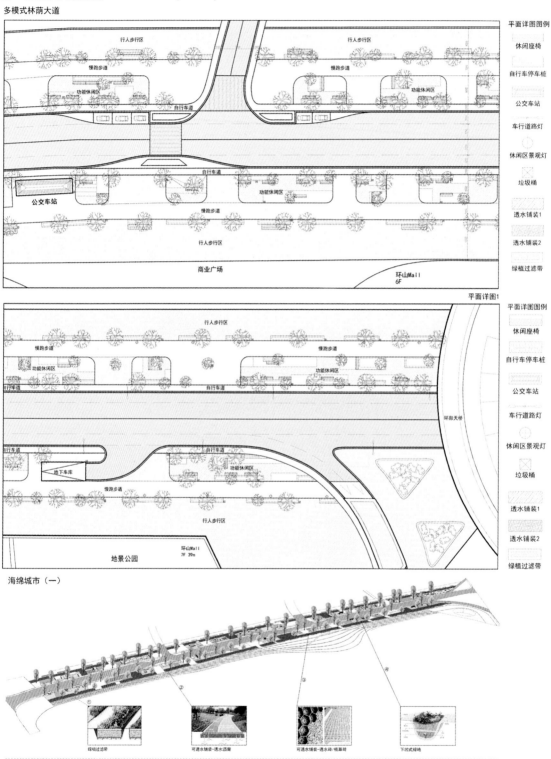

平面详图图例

休闲座椅

自行车停车桩

公交车站

车行道路灯

休闲区景观灯

垃圾桶

透水铺装1

透水铺装2

绿植过滤带

行人步行区

慢跑步道

功能休闲区

自行车道

公交车站

慢跑步道

行人步行区

商业广场

环山Mall
6F

平面详图1

平面详图图例

休闲座椅

自行车停车桩

公交车站

车行道路灯

休闲区景观灯

垃圾桶

透水铺装1

透水铺装2

绿植过滤带

行人步行区

慢跑步道

功能休闲区

自行车道

地下车库

慢跑步道

功能休闲区

行人步行区

地景公园

环山Mall
7F 39m

环翠天桥

海绵城市（一）

绿植过滤带

可透水铺装-透水透青

可透水铺装-透水砖/植草砖

下凹式绿地

260

银河公园

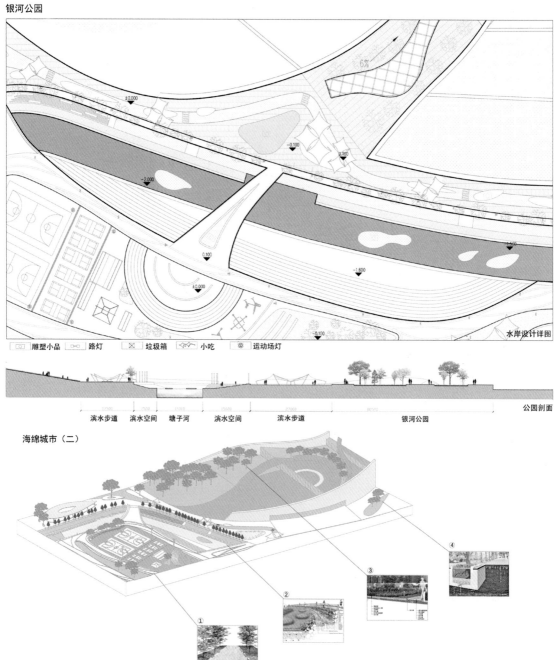

图例：雕塑小品　路灯　垃圾箱　小吃　运动场灯

水岸设计详图

公园剖面

滨水步道　滨水空间　塘子河　滨水空间　滨水步道　银河公园

海绵城市（二）

1. 透水路面和两旁的下沉绿化之间设置有开口的路沿石，便于路面雨水排至绿地。
2. 沿水湿地通过植物的组合来净化雨水径流的水质，河中设置芦苇区深入净化，同时丰富了滨水景观。
3. 屋顶花园有利于收集并利用大面积屋面的雨水，同时能提供舒适的微环境，缓解热岛效应。
4. 在斜坡终点处设置生态树池，收集雨水用于灌溉绿地。

课题五：岳麓山大科城核心区城市设计
Topic 5: Urban design of the core area of Yuelushan Dake City

● **总平面图** General plan

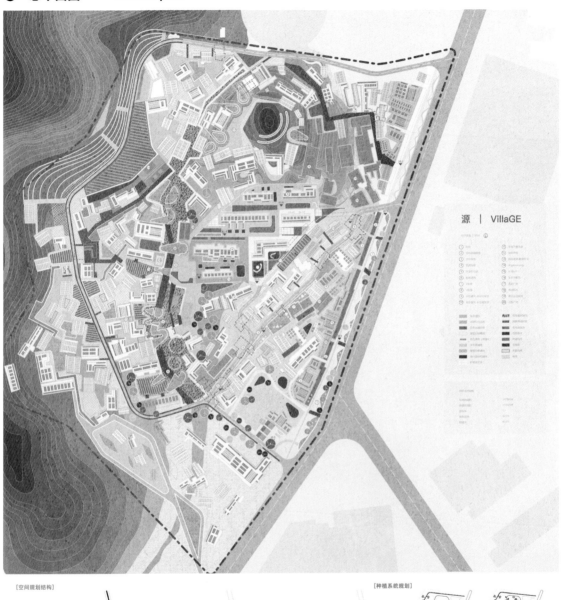

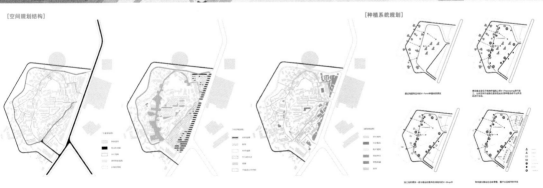

[空间规划结构] [种植系统规划]

設計題目：源
指導老師：蔣甦琦
学　　生：张嘉晟、赵伯轩、徐正文、邓紫玉

Design topic: VillaGE

Instructors: Jiang Suqi

Students: Zhang Jiasheng，Zhao Boxuan，Xu Zhengwen，Deng Ziyu

11

● **前期分析**　　Prophase analysis

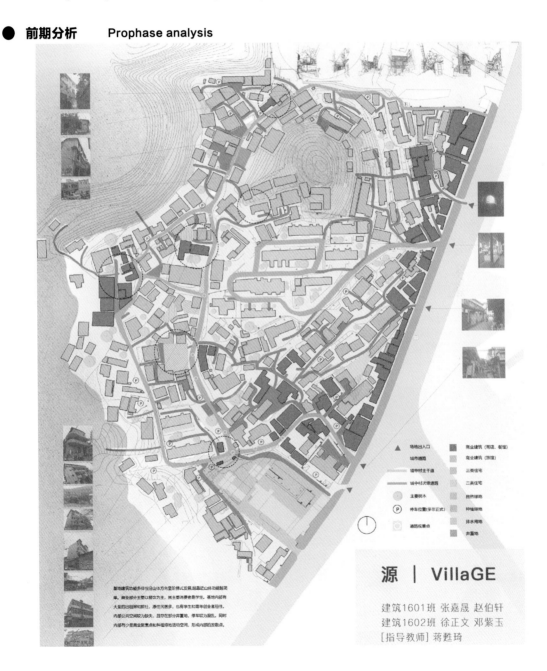

源｜VillaGE

建筑1601班 张嘉晟 赵伯轩
建筑1602班 徐正文 邓紫玉
[指导教师] 蒋甦琦

规划策略   Planning strategy

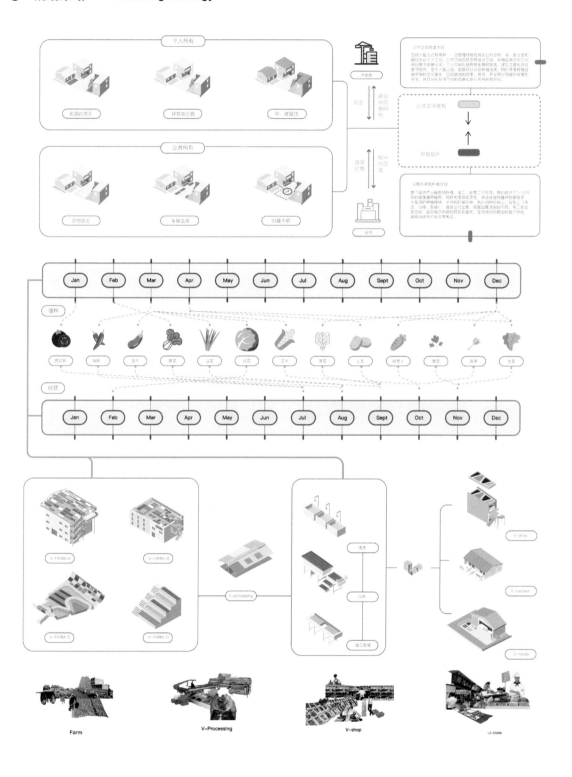

● 节点设计　　Nodes design

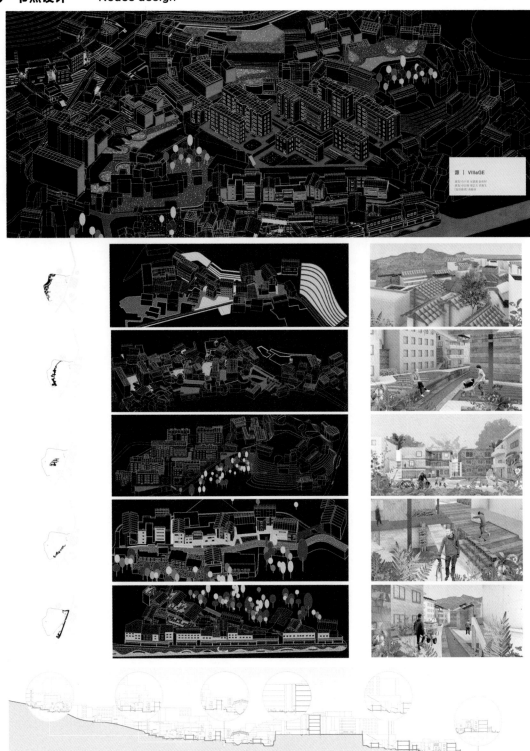

● **总平面图及效果图** General plan and design sketch

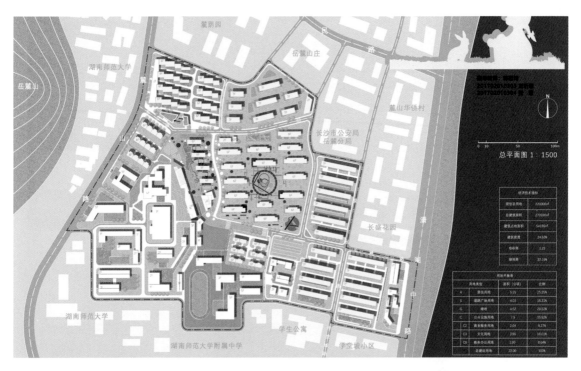

设计题目： 稚趣放学路·乐享街区 —— 儿童友好视角下的儿童成长空间设计
指导老师： 蒋甦琦
学　　生： 刘昕明、张薏

12

Design topic: Children's Growth Space Design from the Children-Friendly Perspective

Instructors: Jiang Suqi

Students: Liu Xinming，Zhang Yi

● **设计策略**　　**Design strategy**

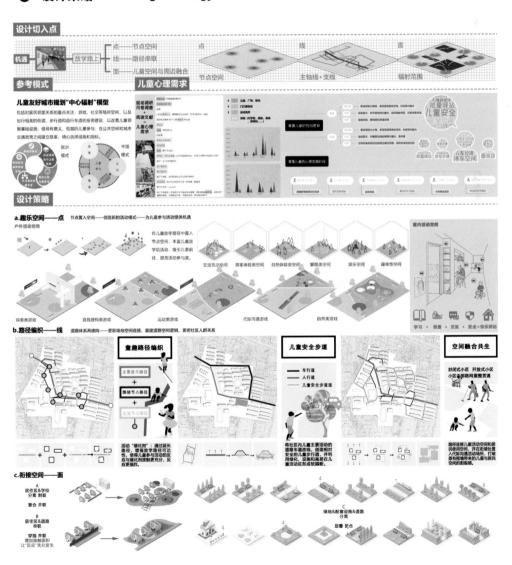

湖南大学建筑与规划学院优秀课程设计汇编
指导老师汇总

二年级教研组　　Grade 2 Teaching and Research Group

叶强
YE Qiang

B.1964, 中国
南京大学博士
湖南大学教授

B. 1964, China
PHD, Nanjing
University
Professor of Hunan
University

谢菲
XIE Fei

B.1973, 中国
英国诺丁汉大学博士
湖南大学副教授

B. 1964, China
PHD, The University of
Nottingham
Associate professor of
Hunan University

杨涛
YANG Tao

B.1988, 中国
美国夏威夷大学博士
湖南大学副教授

B. 1988, China
PHD, University of
Hawaii
Associate professor of
Hunan University

苗欣
MIAO Xin

B.1970, 中国
湖南大学硕士
湖南大学助理教授

B. 1970, China
MArch, Hunan
University
Assistant Professor, Hunan
University

向昊
XIANG Hao

B.1971, 中国
重庆大学硕士
湖南大学助理教授

B. 1971 China
MArch, Chongqing
University
Assistant Professor, Hunan
University

李煦
LI Xu

B.1978, 中国
湖南大学硕士
湖南大学助理教授

B. 1978, China
MArch, Hunan
University
Assistant Professor,
Hunan University

何成
He Cheng

B.1982, 中国
天津大学博士
湖南大学助理教授

B. 1982, China
PHD, Tianjin
University
Assistant Professor, Hunan
University

余燚
YU Yi

B.1984, 中国
都灵理工大学博士
湖南大学助理教授

B. 1984, China
PHD, Politecnico di Torino
Assistant Professor, Hunan
University

李理
LI Li

B.1988, 中国
日本千叶大学博士
湖南大学助理教授

B. 1988, China
PHD, Chiba University
Assistant Professor, Hunan
University

三年级教研组　　Grade 3 Teaching and Research Group

蒋甦琦
JIANG Suqi

B.1969, 中国
湖南大学硕士
湖南大学副教授

B. 1969, China
MArch, Hunan
University
Associate professor of
Hunan University

陈翚
CHEN Hui

B.1971, 中国
捷克技术大学博士
湖南大学副教授

B. 1971, China
PHD, Czech Technical
University
Professor of Hunan
University

李旭
LI Xu

B.1975, 中国
湖南大学博士
湖南大学副教授

B. 1975, China
PHD, Hunan
University
Associate professor of
Hunan University

彭智谋
PENG Zhimou

B.1979, 中国
湖南大学硕士
高级工程师

B. 1964, China
MArch, Hunan University
Senior engineer

张蔚
ZHANG Wei

B.1969, 中国
湖南大学硕士
湖南大学助理教授

B. 1969, China
MArch, Hunan
University
Assistant Professor, Hunan
University

龚震西
GONG Zhenxi

B.1975, 中国
湖南大学硕士
湖南大学助理教授

B. 1975, China
MArch, Hunan
University
Assistant Professor, Hunan
University

罗荩
LUO Jin

B.1975, 中国
湖南大学硕士
湖南大学助理教授

B. 1975, China
MArch, Hunan University
Assistant Professor, Hunan
University

许昊皓
XU Haohao

B.1985, 中国
湖南大学博士
东南大学博士后
湖南大学助理教授

B. 1985, China
PHD, Hunan
University
PD,Southeast University
Assistant Professor, Hunan
University

张光
ZHANG Guang

B.1985，中国
湖南大学博士
湖南大学助理教授

B. 1985, China
PHD, Hunan
University
Assistant Professor, Hunan
University

吕瑞杰
LV Ruijie

B. 1987，中国
香港中文大学博士
湖南大学助理教授

B.1987, China
PhD, The Chinese
University of Hong Kong
Assistant Professor, Hunan
University

四年级教研组　　Grade 4 Teaching and Research Group

王小凡
WANG Xiao Fan

B.1959, 中国
湖南大学硕士
湖南大学教授

B. 1959, China
MArch, Hunan
University
Professor of Hunan
University

袁朝晖
YUAN Chao Hui

B.1970, 中国
湖南大学硕士
湖南大学教授

B. 1970, China
MArch, Hunan
University
Associate professor of
 Hunan University

邓广
DENG Guang

B.1970, 中国
湖南大学硕士
湖南大学教授

B. 1970, China
MArch, Hunan
University
Associate professor of
Hunan University

徐峰
XU Feng

B.1971, 中国
湖南大学硕士
湖南大学教授

B. 1971, China
MArch, Hunan
University
Professor of Hunan
 University

卢健松
LU Jian Song

B.1975, 中国
清华大学博士
湖南大学教授

B. 1964, China
PHD,Tsinghua
University
Professor of Hunan
University

陈晓明
CHEN Xiao Ming

B.1970, 中国
湖南大学硕士
湖南大学副教授

B. 1970, China
MArch, Hunan
University
Associate professor of
Hunan University

齐靖
QI Jing

B.1977, 中国
湖南大学博士
高级工程师

B. 1977, China
PHD, Hunan
University
Senior engineer

宋明星
SONG Ming xing

B.1978, 中国
湖南大学博士
湖南大学副教授

B. 1978, China
PHD, Hunan
University
Associate professor of
Hunan University

严湘琦
YAN Xiang Qi

B.1979, 中国
湖南大学博士
湖南大学副教授

B. 1964, China
PHD, Hunan
University
Associate professor of
Hunan University

刘尔希
LIU Er Xi

B.1975, 中国
湖南大学硕士
湖南大学助理教授

B. 1964, China
MArch, Hunan
University
Assistant Professor, Hunan
University

城市设计教研组　　Urban design Teaching and Research Group

叶强
YE Qiang

B.1964, 中国
南京大学博士
湖南大学教授

B. 1964, China
PHD, Nanjing
University
Professor of Hunan
University

蒋甦琦
JIANG Suqi

B.1969, 中国
湖南大学硕士
湖南大学副教授

B. 1969, China
MArch, Hunan
University
Associate professor of
Hunan University

姜敏
JIANG Min

B.1977, 中国
湖南大学博士
湖南大学副教授

B. 1977, China
PHD, Hunan
University
Associate professor of
Hunan University

陈煊
CHEN Xuan

B.1981, 中国
华中科技大学博士
湖南大学副教授

B. 1981, China
PHD, Huazhong
University of science and
technology
Associate professor of
Hunan University

向辉
XIANG Hui

B.1977, 中国
德国魏玛包豪斯大学硕士
湖南大学助理教授

B. 1977, China
MArch, Bauhaus
University Weimar,
Germany
Assistant Professor, Hunan
University University

彭科
PENG Ke

B.1980, 中国
美国北卡罗来纳大学教堂
山分校博士
湖南大学助理教授

B. 1980, China
PHD,University of North
Carolina Chapel Hill
Assistant Professor, Hunan
University

许昊皓
XU Haohao

B.1985, 中国
湖南大学博士
东南大学博士后
湖南大学助理教授

B. 1985, China
PHD, Hunan
University
PD,Southeast University
Assistant Professor, Hunan
University

孙亮
SUN Liang

B.1989, 中国
华中科技大学博士
湖南大学助理教授

B. 1989, China
PHD, Huazhong
University of science and
technology
Assistant Professor, Hunan
University
 University